Praise for Tim Houlne and *The Intelligent Workforce*

The Intelligent Workforce addresses the future, where the pace of AI and automation continues to increase, as well as the new job opportunities it creates, as only humans have the ability to fully translate insights from the digital to the real world.

MARC BUNCHER, *President and CEO, Siemens Mobility*

Tim and *The Intelligent Workforce* are on the forefront of change and what it means for your business. An early adopter of the cloud and remote work, he now advocates for embracing AI and machine learning collaborating with human intuition as perhaps one of the biggest challenges and benefits business and society has ever faced.

DORY WILEY, *President and CEO, Commerce Street Holdings*

The Intelligent Workforce explores the transformative relationship between human creativity and machine intelligence that is reshaping our workplaces and industries. This book serves as a crucial guide for anyone eager to understand and harness the potential of this dynamic collaboration in the era of technological revolution.

STEVE ROWLEY, *President, Cox Automotive*

The Intelligent Workforce lays out the facts of the relationship between humans and technology (AI, GenAi, etc.) and guides the reader through the opportunities for collaboration and augmentation and the risks of misuse and replacement. Ultimately, it will be up to all of us to understand how we can leverage AI and GenAI to cocreate a better future.

BARBARA PORTER, *Managing Director, CX Practice Leader, EY Consulting*

"In *The Intelligent Workforce*, Tim lays a solid foundation and road map for AI use, advocating a practical approach for human oversight. His pioneering insights demystify AI, reassuring those wary of machine displacement. Drawing from his pioneering experience as an early innovator in virtual workforces and their supporting applications, Tim brings valuable insights into the technologies influencing customer service."

CHRIS SMITH, *Chief Service Avenger, OnPoint Warranty Solutions*

THE
INTELLIGENT
WORKFORCE

TIM HOULNE

THE
INTELLIGENT
WORKFORCE

HOW **HUMANS + MACHINES** WILL
CO-CREATE A BETTER FUTURE

Forbes | Books

Published by Forbes Books, Charleston, South Carolina.
An imprint of Advantage Media Group.

Forbes Books is a registered trademark, and the Forbes Books colophon is a trademark of Forbes Media, LLC.

Printed in the United States of America.

10 9 8 7 6 5 4 3 2 1

ISBN: 979-8-88750-160-4 (Hardcover)
ISBN: 979-8-88750-161-1 (eBook)

Library of Congress Control Number: 2024908949

Cover design by Megan Elger.
Layout design by Wesley Strickland.

This custom publication is intended to provide accurate information and the opinions of the author in regard to the subject matter covered. It is sold with the understanding that the publisher, Forbes Books, is not engaged in rendering legal, financial, or professional services of any kind. If legal advice or other expert assistance is required, the reader is advised to seek the services of a competent professional.

Since 1917, Forbes has remained steadfast in its mission to serve as the defining voice of entrepreneurial capitalism. Forbes Books, launched in 2016 through a partnership with Advantage Media, furthers that aim by helping business and thought leaders bring their stories, passion, and knowledge to the forefront in custom books. Opinions expressed by Forbes Books authors are their own. To be considered for publication, please visit **books.Forbes.com**.

To all forward-thinking executives at innovative companies, to the curious minds interested in the intersection of humans and technology, and to all those who believe in the power of collaboration between humans and machines. This book is dedicated to you. May this book inspire you to embrace the concept of an intelligent workforce and leverage the power of intangible and tangible AI to grow your careers and businesses.

CONTENTS

ACKNOWLEDGMENTS

THANK YOU TO FORBES BOOKS AND ADVANTAGE MEDIA GROUP, the editorial team including Lauren Steffes, the editorial manager, and collaborative writer Ezra Byer for your invaluable feedback, guidance, and support throughout the writing and editing process.

I also want to thank my wife, Kim, and my sons, Nick and Jack, for their patience, understanding, and encouragement during the writing of the book.

To the forward-thinking executives at innovative companies, the target audience of the book, I value your willingness to embrace new technologies and drive the future of work.

To you, the reader, thank you for your interest and engagement in the subject matter and for your commitment to understanding and navigating the future of the intelligent workforce.

And to all my colleagues and mentors who have contributed to my understanding and perspective on the intersection of humans and technology, you have my appreciation.

Special thanks to all who provided testimonials or quotes on how AI is changing the way business works, including Dr. John Hansen, Gary Ash, Ryan Chavez, Marc Buncher, Dan Gingiss, Scott Hermann, Dennis Wakabayashi, Damien Harmon, Mike Mateer, Michael Saracini, Doug Ellison, Steve Rowley, Barbara Porter, Chris Smith, Mason Levy, and Aaron Bergman.

Finally, thank you to the companies and educational institutions that are pioneering the use of AI and automation in their operations, providing valuable case studies and insights for this book.

INTRODUCTION

WELCOME TO A JOURNEY INTO THE FUTURE OF WORK, A WORLD where humans and machines coexist and collaborate, creating an intelligent workforce that is more than the sum of its parts. This book is a guide to understanding and navigating this new landscape, where artificial intelligence (AI) and human intelligence blend to create unprecedented opportunities for businesses and individuals alike.

From the earliest tools wielded by our ancestors to the latest advancements in artificial intelligence and robotics, humanity has always sought to augment its capabilities. In the era of digital transformation, the intelligent workforce is not a distant dream but a rapidly evolving reality. It's a world where AI does not replace humans but makes them more valuable, augmenting their capabilities and freeing them from mundane tasks. This allows humans to focus on what they do best—applying emotional intelligence, making creative decisions, and solving complex problems.

The intelligent workforce is a symphony of intellects, where cognitive human capabilities and the analytical prowess of machines meld together, creating an intelligent workforce unparalleled in history. It represents an evolution, a blend of emotional intelligence, creative problem-solving, relentless efficiency, and analytical rigor that propels organizations into new frontiers of problem-solving and value creation. Humans contribute empathetic, creative, and strategic thinking; machines bring precision, scalability, and data-driven insights, together shaping a future where the collective intelligence is greater than the sum of its parts.

However, this transformation is not without its challenges. Here we will address the subtle art of balancing technological advancements with ethical considerations, ensuring the harmonization of artificial and human intelligence yields benefits that are not merely economic but profoundly social and humanitarian.

As we move toward a future where AI and humans work side by side, it is crucial to understand the implications of this shift.

> ♀ How do we ensure a balance between efficiency and ethics?
>
> ♀ How do we prepare our workforce for this new reality?
>
> ♀ And most importantly, how do we use this technology to enhance human potential rather than diminish it?

In this book, we will embark on a journey through the framework of the intelligent workforce, exploring the dynamics of how humans and machines, with their respective strengths and vulnerabilities, converge to redefine the parameters of productivity, creativity, and innovation.

Envision a world where technological entities do not usurp but uplift—where machines do not replace but rather empower. Within these chapters lies a journey through diverse industries and multifaceted roles, exploring how this alliance between human and machine intelligence is quietly rewriting the rule book of operational excellence and strategic innovation.

So whether you are a senior leader looking for strategies needed to develop contextual intelligence to grow your business, a CEO looking to future-proof your organization, an employee navigating the changing job market, or simply a curious mind interested in the intersection of humans and technology, this book is for you.

Let's embark on this journey together, toward a future where humans and machines collaborate to create a truly intelligent workforce. This book will help with your justification to invest in the technologies and skill sets of tomorrow for augmented intelligence, where the best outcomes are achieved when humans and machines work together.

CHAPTER 1:

EVERYTHING IS DIGITAL

EVERYTHING IN BUSINESS AND LIFE IS NOW DIGITAL, AND THE lines between the physical and virtual continue to blur at an unfathomable rate.

The new world of work has fundamentally shifted how businesses operate—providing employees with a real-time, constant stream of emails, messages, and digital communications. In a report posted by Canada-based company Deloitte, researchers observed that "as the distinction between professional and personal life dissolves, and the workplace becomes truly digital, employees are communicating and collaborating in unprecedented ways."[1]

Social posting and real-time updates have virtually no boundaries or geographical restrictions.

> ♀ What should I post today?
>
> ♀ What platform should I use to convey my message?

> ♀ Should I tell people what I had for dinner?
>
> ♀ Do I write about where my family is vacationing this week?

These are the questions social media users wrestle with each day. Common information individuals previously wouldn't have considered sharing with people outside their close social circle they now freely distribute to hundreds, thousands, and even millions of people. Often, these updates are not limited to one social network circle of influence, and a message shared on one channel can be shared with the friends of numerous networks.

This same form of social influence is now applicable to work. In a report published by PwC, researchers concluded the following:

> We are living through a fundamental transformation in the way we work. Automation and "thinking machines" are replacing human tasks and jobs and changing the skills that organizations are looking for in their people. These momentous changes raise huge organizational, talent and HR challenges—at a time when business leaders are already wrestling with unprecedented risks, disruption and political and societal upheaval.[2]

Consider for a moment the rise of digital technology in the past thirty years. Unlike previous generations that saw little technological or innovative advancement from one generation to the next, the last few decades have brought our world closer together than ever before. Video conversations with virtually any person of our choosing can take place around the globe. And this is just the beginning.

Adapt or Die

Over a decade ago, I wrote in my first edition of *The New World of Work* that we were undertaking a massive transition from the cube to the cloud. Business wouldn't just be conducted in the office, and workers could work remotely. I shared how it was my belief that in the next few years, the workforce would shift and give way to a new generation of entrepreneurs who saw the value of having teams work in remote locations.

Admittedly, this declaration was met with much resistance. Working from home "doesn't work for young kids or spontaneity or management,"[3] said JPMorgan chief Jamie Dimon at Davos in 2023.

But two years of the COVID-19 pandemic revealed just how misplaced these criticisms were. To the surprise of many business owners, the forced shift to remote work increased company productivity.[4] The pandemic changed the way we looked at established realities of business, such as working in an office. With many shifting to virtual workspaces, the need for physical locations became unnecessary. The same technology that blended our professional and personal lives now allows us to do our jobs from home or anywhere in the world.

The problem is that humans can be stubborn creatures of habit. We like to do and live in ways that feel comfortable. Few leaders thought they would live to see the day when most of their team did not come into an office from nine to five. But for many, this is precisely what has happened. They did not welcome this shift, but it was thrust upon them, and they faced that age-old conundrum: adapt or die.

As I've stated for years, work has been fractionalized, talent has been globalized. and technology has been virtualized, paving the way for a new world of work—a digital world shared with real-world artificial intelligence. And contrary to what many predict, AI, robotics, and automation are not going to destroy jobs. Instead, they are going to create new opportunities—the same way technology has done with every wave of innovation. There is a massive labor shortage today, and real-world AI in a digital world will be a welcome addition to the new digital workforce.

As life-changing as this cube-to-cloud revolution was, there is an even greater shift entrepreneurs of the next decade will face. It will dramatically impact the way they live and do business, and their ability to adapt to this digital change will determine whether their businesses survive or die.

A New Model for Work

Many executives think owning smartphones or moving away from physical files to digital means they are current with technological trends. However, entrepreneurs who plan to be in business for the long term must understand that the level of human and artificial intelligence interactions will increase at an even more rapid rate than the digital transitions of the past.

In 2023, it's estimated that just over 85 percent of people in the world own smartphones.[5] Just as a snowball gains speed and size when it rolls down a hill, the digital trends of tomorrow will far surpass the speed and size of digital trends of the past. Artificial intelligence, automation, and robotic process automation are taking the place of traditional jobs at an unprecedented rate.

"The pace of change is accelerating," researchers conclude. "Competition for the right talent is fierce. And 'talent' no longer means the same as ten years ago; many of the roles, skills and job titles of tomorrow are unknown to us today."[6] On the other hand, jobs that are not lost now have digital assistance from AI and automation, improving deliverables, timeframes, and outcomes.

This new wave means large commercial leases will shrink, giving way to a rise in individuals who work from home. In fact, we're already seeing a more blended model of the physical and home office, with many commuting to work for a few days and working from home the remainder of the week. With the increased demand for specialized talent, employers realize it is not feasible to limit their search for quality job candidates to employees living within a thirty-to-forty-mile radius of their office.

Add to this wave the growth of the gig economy, and we see a flexible workforce that can quickly adapt to changing trends and patterns of consumers. Back in 2017, *Forbes* projected the gig economy would surpass the full-time workforce by 2027.[7] The COVID-19 pandemic only accelerated this estimation. From delivery drivers to freelance editors, large corporations and small companies alike can outsource talent to competent individuals who are skilled in a particular craft. This, in turn, means they do not need to spend as much on a full-time workforce and can work with individuals who do not need to divide their expertise in multiple areas. In an article titled "Workforce Strategies in the Era of the Gig Economy," the authors provide this statement of cautious optimism:

> There's no question, though, that flexible labor will become a more entrenched part of the workforce. This will require organizations to think more strategically about how work is done. Delivering a good experience to temporary workers increases their loyalty to the company, helps ensure that they will be available when needed, and reduces onboarding times. Organizations that crack the code will be able to tap into

these growing ecosystems faster and more effectively than their competitors.[8]

For managers, it is critical they adapt to this digital transformation and job reset. Work is changing, careers are shifting, and those long-sought-after titles and levels of seniority once thought invaluable seemingly devalue by the day, leaving many to conclude, "successful companies will make themselves a place where short-term and temporary workers want to work."[9]

With this shift in global digitization, every individual now has a platform to share their views of the world. Few who fall into the Generation Z category are content to join an organization, show up each day to offer their contributions, and have little say concerning the ins and outs of how their departments are run. Instead, they want to be heard and have a seat at the table. And it is up to employers and leaders of organizations to adjust to this way of thinking.

Workers today want more than just a paycheck. They want a solid work-and-life balance. And if employers fail to deliver on their expectations, workers will leave and monetize their talents elsewhere. But here is the good news: AI can help improve work processes, thus making for a better work experience and creating an environment that retains skilled workers. Tasks can be completed faster, more efficiently, and with improved accuracy.

In short, for most organizations to survive in the coming decade, they must recognize the emerging patterns of the digital workforce, understand the impact these changes will have on their teams and customers, and then take proactive steps to work *with* rather than *against* these changes. Just as a surfer cannot stop the waves in an ocean, so a business owner in this digital era cannot stand against the digital tsunami headed their way. Instead of pushing against the waves, leaders must learn to adapt, ride the waves, and use them to their advantage.

Benefits and Shortcomings

This technological shift comes with its obvious benefits and short-comings. There will always be jobs that humans alone can do. These include roles that require a moral compass and a nuanced understanding of the way the world works. As many have noted, "The future of work asks us to consider the biggest questions of our age."[10]

But setting these risks aside, we find the benefits and potential of technology are tremendous. Technology is replacing and will continue to replace menial tasks such as stocking shelves, driving, and working in a call center. Yes, in the short term, technology might eliminate basic job roles. But in the long term, AI always creates fresh opportunities and requires operators to oversee its progress.

Some have embraced this shift. But others are more hesitant. A 2020 study of businesses conducted by McKinsey made this observation:

> Overall, half of respondents say their organizations have adopted AI in at least one function. And while AI adoption was about equal across regions last year, this year's respondents working for companies with headquarters in Latin American countries and in other developing countries are much less likely than those elsewhere to report that their companies have embedded AI into a process or product in at least one function or business unit. By industry, respondents in the high-tech and telecom sectors are again the most likely to report AI adoption, with the automotive and assembly sector falling just behind them (down from sharing the lead last year).[11]

The benefits are great. But some would argue they come with too great a price tag. For many organizations, the days of catching up with coworkers at the watercooler are things of the past as in-person interactions become less and less frequent. Is this a good thing?

The answer to this question is challenging. On one hand, the advancements that come with this change are beneficial. But if there was anything we learned from the COVID-19 pandemic, face-to-face communication is still important. In the words of Microsoft CEO, Satya Nadella, "Digital technology should not be a substitute for human connection."[12]

This same McKinsey Global Survey noted above revealed "that organizations are using AI as a tool for generating value."[13] And this value is coming through revenue. The article continues by saying, "A small contingent of respondents coming from a variety of industries attribute 20 percent or more of their organizations' earnings before interest and taxes (EBIT) to AI."[14]

This survey went on to note, "These companies plan to invest even more in AI in response to the COVID-19 pandemic and its acceleration of all things digital." But this advancement comes with a word of caution, as it suggests the following:

> This could create a wider divide between AI leaders and the majority of companies still struggling to capitalize on the technology; however, these leaders engage in a number of practices that could offer helpful hints for success. And while companies overall are making some progress in mitigating the risks of AI, most still have a long way to go.[15]

The implications are clear. Businesses that continue to invest in AI will accelerate and widen the gap between those that do not. Where the "full saturation" point occurs and what the peak of human and

AI efficiency and productivity might be remain a mystery. But in this rise and growth of AI, we must do all we can to guard against the burnout that will naturally occur with those who struggle to keep up with the changing times.

The rise of AI will make most people better off over the next few decades, but how will those who fail to adapt evolve and survive in a future world? What solutions can be put in place to address the negative impacts on humanity? And how should these issues be addressed with the same effort as the innovation?

As expert data analyst Susan Etlinger notes, "In order for AI technologies to be truly transformative in a positive way, we need a set of ethical norms, standards and practical methodologies to ensure that we use AI responsibly and to the benefit of humanity."[16]

The Rise of Artificial Intelligence

If you've never watched the documentary *AlphaGo*, you need to. In it, American filmmaker Greg Kohs tells the story of how a team of artificial intelligence researchers at DeepMind, a British-American artificial intelligence research laboratory, were able to craft a self-learning program that was able to beat the reigning world champion of the popular age-old board game Go. Unlike the Deep Blue program that beat Gary Kasporov in 1997 by having a team of professional chess players insert their knowledge of the game, the artificial intelligence used in this instance was a machine learning program that continued to get better and better with time.

At its core, Go is a simple game that has the basic objective of claiming as much territory on a board as possible. But while the premise of the game is easy to grasp, the almost infinite number of positions that can be played make it impossible for the human mind to master. In fact, the number of configurations on a Go board are more than the number of atoms in the entire universe.[17] As one of the founders of DeepMind, Demis Hassabis, stated, "Even if you took all the computers in the world and ran them for a million years that wouldn't be enough compute power to calculate all the possible variations."[18]

This was the ultimate test for AI advancement. Virtually everyone thought AI was decades away from beating the best minds in Go. But in October of 2015, AlphaGo defeated the European champion Fan Hui in five straight matches, proving the incredible advancement AI had made. Still, the world was skeptical and doubted a program like AlphaGo could compete with the best player in the world at that time in Lee Sedol, who was rated seven levels higher than Hui.

Five matches were set in Seoul, South Korea, between March 9 and March 15, 2016. Coming into this series, Sedol made no secret that he intended to win all five games with ease. Many experts made the same observation. But Sedol's confidence was quickly shattered as AlphaGo gave a dominant performance in game 1 and took an early lead in the best-of-five series. This loss prompted Gary Kasparov, a man who resonated with Sedol's pain, to tweet, "Condolences to Lee Se-dol on losing game one to AlphaGo. I hope he can recover, but the writing is on the wall."[19] Kasparov's prediction proved accurate, and Sedol won only one of the remaining four matches with AlphaGo.

As you watch this documentary, you cannot help but be gripped by the mental, physical, and emotional weight Sedol felt with seemingly the hope of humanity resting on his shoulders. After losing three straight games, you could see the pain in Sedol's expression. But something curious happened in game 2 that became a focal point of the way Sedol viewed these matches.

AlphaGo played a seemingly poor move that its own calculations estimated only one in ten thousand humans would attempt. Most commentators thought it was a drastic miscalculation. But as it turned out, this move proved to be a masterful stroke of genius and shifted the balance of the match. Reflecting on this move after the match, Sedol made this keen observation: "I thought AlphaGo was based on probability calculation, and it was merely a machine. But when I saw this move, I changed my mind. Surely AlphaGo is creative."[20]

Sedol realized this form of AI could play with imagination and go beyond preset human capabilities. While he lost the next game, this remarkable move by AlphaGo challenged the way he viewed the game. And in game 4, Sedol played what was termed a "wedge move" on move 78 that finally found a blind spot in AlphaGo's system. It was the turning point, and Sedol went on to win that match, giving hope to many around the world that humans could still compete in a world of increased AI. This left one of the commentators of the match to remark,

> At least in a broad sense, move 37 begat move 78 [the critical move in Game 4 that allowed Lee Sedol to win] and begat a new attitude to Lee Sedol a new way of seeing the game. He improved through this machine. His humanness was expanded after playing this inanimate creation. And the hope is that that machine, and in particular the technology behind it can have the same effect with all of us.[21]

AlphaGo was only the beginning. On November 30, 2022, the world responded to the release of ChatGPT. This marked a digital renaissance that fundamentally changed the world's view on generative AI. In the vast corridors of the digital cosmos, there emerged a beacon, a source of wisdom. While it once seemed impossible that any company would unseat Google as the search engine standard or challenge it for superiority regarding artificial intelligence, since its launch, ChatGPT has become increasingly popular across multiple industries. It rocketed to over one hundred million users in just three months between February 2023 and April 2023, and it hit its first million users in only five days, a feat that took Netflix 3.5 years to achieve.

ChatGPT was built on the GPT family of large language models (LLMs) and delivers detailed responses and articulate answers across many knowledge domains. Its chatbot structure and generative AI's futuristic capability allow users to quickly generate content based on a variety of inputs: "ChatGPT's dense 'neural' network consists of over 175 billion parameters and incredible natural language processing (NLP) abilities that perform tasks with only a few lines of input—offering a flashy tool everyone wants to use."[22]

Now users are able to perform tasks once thought unimaginable.

Our Relationship with AI

After his loss to Deep Blue in 1997, Gary Kasparov had time to reflect on this experience and made this statement: "A good human plus a machine is the best combination."[23]

It's important to develop a healthy relationship with AI. Human and artificial intelligence are not mutually independent and do not operate in a vacuum. The best outcomes are achieved when humans and machines work together. This is called augmented intelligence, and business leaders need to compel every level of management to learn how to augment employee performance with digital technology.

AI will not replace humans. Rather, it will make humans more valuable. Humans have an emotional intelligence that machines are not able to match. AI can interpret millions of data variables, but we still need to interpret the decision with a human. Humans are still required to tune the machine for editorial or ethical content, intent, and real-life sentiment analysis. However, over time, guided by human input, machines will be able to compute real-life scenarios.

Unlike humans, AI does not have emotion. This can be both good and bad, and as more of the workforce is controlled by digital tools, it is imperative these advancements have the elements of a "human touch." In the article "Workforce of the Future," the writers argue it is helpful to think of three levels of AI:

> **Assisted intelligence**, widely available today, improves what people and organizations are already doing. A simple example, prevalent in cars today, is the GPS navigation program that offers directions to drivers and adjusts to road conditions.
>
> **Augmented intelligence**, emerging today, helps people and organizations to do things they couldn't otherwise do. For example, car ride-sharing businesses couldn't exist without the combination of programs that organize the service.
>
> **Autonomous intelligence**, being developed for the future, establishes machines that act on their own. An example of this will be self-driving vehicles, when they come into widespread use.[24]

If they have not already, these three forms of AI will have massive implications for your business in the coming years. In 2021, Facebook rebranded itself as Meta. And much has been said and written about the metaverse in which many individuals "reside." In the metaverse, inhabitants can purchase NFTs (nonfungible tokens) that allow them to buy virtual items of clothing or objects for their virtual homes. It is a world within a world. Devices such as the Oculus Quest from

Facebook that allow gamers and workout enthusiasts to enter a virtual realm are just the tip of the iceberg.

As authors Amy Colbert, Nick Yee, and Gerard George note,

> We are at the beginning of an exciting transformation of work, work practices, and workplaces. The digital competencies of the workforce and the ways in which technology are used in the workplace will continue to develop and change. This provides organizations and managers with a wealth of possibilities for increasing organizational effectiveness.[25]

In many respects, we have entered a fourth major revolution. The Industrial Revolution of the eighteenth and nineteenth centuries harnessed the power of coal and gas to forever change the way humans interacted with machines. The mass production revolution took this to a new level. Business leaders like William Edwards Deming emphasized the importance of continuous improvement and demonstrated that mass production could be conducted with quality and precision. And the information age revolution saw leaders like Steve Jobs and Bill Gates bring the world together through digital communication.

But this new "machine learning revolution" phase we have entered, which has made language models like ChatGPT possible, is earth shattering. The ability of machines to learn from their mistakes and mimic the neuro-understanding of a human being is a complete game changer. And we as humans must understand the impact digitization will have on our lives and the companies we lead. Almost every aspect of our lives is impacted in some way by AI, but it's up to us to understand this revolution and harness its power to our advantage.

Predictions and Observations

Considering the digital technology outlook and ongoing trajectory of newer innovations for a future intelligent workforce, here are some predictions and observations regarding the intelligent workforce of the future:

- **Continued automation and AI integration.** Jobs that involve repetitive tasks will increasingly be automated. But it's not just about replacement; it's about augmentation. AI will be used to enhance the capabilities of human workers, making them more efficient and informed.

- **Reskilling and upskilling.** With the shift in job roles due to automation and AI, there will be a significant emphasis on reskilling and upskilling the current workforce. Continuous learning will become an integral part of professional life.

- **Remote and flexible work.** The pandemic accelerated the move toward remote work. With advanced collaboration tools and virtual reality, remote work might become even more prevalent, seamless, and efficient.

- **Collaborative robots.** Robots, or "cobots," will work alongside humans in various industries, especially in manufacturing and healthcare, aiding in tasks and ensuring safety.

- **Data literacy.** As businesses become more data driven, there will be a growing demand for employees who can interpret and make decisions based on data. This doesn't mean everyone will be a data scientist, but a basic understanding of data will be invaluable.

- **Emphasis on soft skills.** While technical skills are essential, soft skills like empathy, communication, critical thinking, and problem-solving will be even more critical. As machines take on more technical tasks, human-centric skills will differentiate employees.

- **Digital twins and AR.** The use of digital twins—virtual replicas of physical devices—will grow. Alongside this, augmented reality (AR) will aid in tasks like training, maintenance, and design.

- **Personalized work experiences.** AI will enable more personalized work experiences. For instance, learning and development might be tailored to an individual's career trajectory and personal strengths.

- **Decentralized and edge computing.** As devices get smarter, there's a push toward processing data at the source rather than in a centralized data center—known as edge computing. This will have implications for various industries, from manufacturing to healthcare.

- **Enhanced cybersecurity.** As the digital realm expands, so does the threat landscape. AI and machine learning will play a critical role in predicting, identifying, and countering cyberthreats. However, there'll also be a need for a human touch, necessitating a skilled cybersecurity workforce.

- **Health and well-being.** With wearable tech and advanced biometrics, companies might play a more active role in monitoring and promoting the health and well-being of their employees, especially in physically demanding roles.

- **Sustainability and green tech.** As the world grapples with climate change, there will be a push toward sustainable practices. This will influence job roles, with a rise in green-tech jobs and a focus on sustainable practices in all industries.

- **Diverse and inclusive workforce.** Technology will aid in eliminating biases in hiring processes, ensuring a more diverse and inclusive workforce. This not only promotes fairness but also drives innovation.

💡 **Lifelong learning platforms.** As the pace of technological change accelerates, there will be a rise in platforms that support lifelong learning, enabling professionals to stay updated with the latest skills and knowledge.

The future of work will be marked by a symbiotic relationship between humans and technology. While technology will augment capabilities, the uniquely human traits will remain invaluable. The challenge for organizations will be in balancing this relationship, ensuring that technology empowers rather than replaces the human workforce.

CHAPTER 2:

THE RISE OF DIGITIZATION

THE RISE OF DIGITIZATION HAS BEEN BOTH GRADUAL AND RAPID.

As Harry Menear of *Technology Magazine* notes, "The history of digital transformation goes back further than you may expect."[26] For example, the abacus, one of the earliest counting tools, was used in ancient civilizations. And the concept of digitization has its roots as far back as 1679, when Gottfried Wilhelm Leibniz developed the first-ever binary system. A binary describes a numbering scheme in which there are only two possible values for each digit, 0 or 1, and is the basis for all binary code used in computing systems.

By the seventeenth century, we had mechanical devices like Blaise Pascal's "Pascaline," and Gottfried Wilhelm Leibniz's "Stepped Reckoner," designed for arithmetic operations. The nineteenth century brought us the introduction of punched cards, initially used for weaving patterns in the textile industry and later adapted for data processing and storage. The invention of the telegraph in the 1830s and the telephone in the 1870s marked significant steps in the transmission of information over long distances.

By the early twentieth century, radio began transmitting voice and music, and by midcentury, television began broadcasting moving images and sound—albeit all in analog format. The 1940s saw the birth of digital computers. Then, the invention of the transistor in the late 1940s and the integrated circuit in the late 1950s set the stage for miniaturization and the exponential growth in computing power. The late 1970s and early 1980s marked the shift from analog to digital recording, most notably with the introduction of the compact disc (CD) for music.

The 1980s brought about a revolution in personal computing with the launch of PCs like the IBM Personal Computer and Apple's Macintosh. Although the internet's foundational technologies were developed in the 1960s, it was the creation of the World Wide Web in the early 1990s by Tim Berners-Lee that made the internet accessible to the general public.

The late twentieth century marked the shift from film to digital photography with the launch of consumer digital cameras. In the 2000s, we saw a boom in mobile technology, notably with smartphones. Devices like Apple's iPhone and iPad, and those running Android, merged computing, media, and telecommunication. Apple's iPhone, launched in 2007, was pivotal in this transition.

Platforms like Facebook, Twitter, and YouTube, founded in the mid-2000s, changed the way content was created, shared, and consumed, further driving digitization. By the 2010s, cloud services began to replace local storage and computing, allowing users to access their data and applications from any device with an internet connection. Companies like Amazon, Google, and Microsoft offered cloud services, allowing data storage and processing to move off personal devices.

The abundance of analytics allowed organizations to begin harnessing large datasets, utilizing sophisticated algorithms to gain insights and drive decision-making. Embedding digital technology in everyday objects, from refrigerators to thermostats, has allowed for interconnected ecosystems and smart environments. This relentless pace of technological advancement suggests that the story of digitization will continue at an accelerated pace.

The future of digitization heavily depends on a company's ability to adapt digital systems and a flexible, agile digital infrastructure to cope with competition from other similar firms. Regardless of their size, businesses have been advised to brace for significant changes in the way they do business. If they haven't already, companies will have to move too fast to adapt to business models leveraging AI.

In the words of Jeff Bezos, "There is no alternative to digital transformation. Visionary companies will carve out new strategic options for themselves—those that don't adapt will fail."[27]

The Evolution of Work

From the farm to the factory, to the cube, to the cloud sequences, the transformative journey of work now incorporates computing and data management, tracing the transition from traditional on-premises systems (the cube) to the adoption of cloud computing technologies. The phrase "from the cube to the cloud" refers to the shift in the way work is done—from traditional office spaces to digital—and examines the implications, benefits, challenges, and prospects of this paradigm shift.

In the traditional model, employees would work in physical office spaces, using local hardware and software. However, with the advent of cloud computing, work has become more virtual and location independent. Employees can now work remotely from anywhere in the world, accessing company resources and data stored in the cloud. This shift has led to increased flexibility and efficiency and has opened new possibilities for how businesses operate and how employees work.

This transition also involves a change in mindset, with businesses needing to adapt to new ways of working and to leverage the opportunities provided by digital technologies. It's not just about using new tools but also about changing business processes and models to fully take advantage of the possibilities offered by the cloud.

As I write in the second edition of *The New World of Work*, "The pandemic forced us to completely rethink work, technology, and the ways we connect."[28] Many companies adopted remote or hybrid work arrangements during the pandemic. This flexibility proved to be popular among workers, and many businesses have since decided to make remote or hybrid work a permanent option. Tools like Zoom, Microsoft Teams, and Slack saw increased usage to facilitate collaboration and communication.

Necessity pushed many companies to accelerate their digital transformation. This included the adoption of cloud technologies, automation, and other digital tools to support remote work and maintain operational efficiency. Organizations realized the importance of their employees' mental and physical health. Many started offering wellness programs, mental health resources, and more flexible time-off policies. The concept of "Zoom fatigue" also highlighted the need for breaks and boundaries in a remote work environment.

The pandemic-induced economic downturn forced many businesses to lay off or furlough workers. However, sectors like e-commerce, healthcare, and tech experienced growth, leading to a reallocation of labor. The shift in job roles and responsibilities prompted a greater emphasis on upskilling and reskilling. Online learning platforms like Coursera, Udemy, and LinkedIn Learning all increased enrollments.

As companies adopted hybrid work models, the design and purpose of office spaces began to change. Some businesses reduced their physical office footprints, while others redesigned their spaces to be more collaborative and flexible. The rise of virtual meetings and events significantly reduced business travel. With the emphasis on remote work, some companies began offering home office stipends, allowances for internet connectivity, and even stipends for coworking spaces. Companies became more aware of the importance of having a crisis response plan, having experienced the unprecedented disruptions caused by the pandemic.

The pandemic, coupled with global social justice movements, highlighted disparities in healthcare, job security, and other areas. This intensified the push for DEI initiatives in the workplace. Employees began valuing job security, work-life balance, and purpose-driven work more than before. This has implications for employer branding and recruitment strategies.

These changes, while triggered by a crisis, are shaping the future of work in ways that many believe will be more resilient, flexible, and inclusive. However, challenges like maintaining company culture, addressing inequalities, and ensuring effective communication in remote settings remain areas of focus for many businesses.

Expansion in Computing Power

Digitization changes everything. As European Investment Bank editor Dawid Fusiek writes, "Digitalisation speeds up development, helps economic growth, brings people closer together and enables better use of resources."[29]

In the dawn of the twentieth century, machines hummed and clicked, processing data through punched cards in vast rooms. These machines, though groundbreaking for their time, were rudimentary and specialized. The concept of a "computer" as we understand it today was a distant dream. Enter the 1940s, and with them, the birth of the electronic computer. The ENIAC, often touted as the first general-purpose electronic computer, changed the course of technological history. No longer bound by mechanical constraints, these electronic marvels could process complex calculations at speeds once deemed impossible. The torch had been lit.

By the 1950s and '60s, transistors replaced vacuum tubes, leading to smaller, more reliable, and powerful machines. This miniaturization set the stage for the silicon revolution of the 1970s. Integrated circuits, with their minute transistors, exponentially amplified computational capabilities. Moore's Law, the observation that the number of transistors on a microchip doubles approximately every two years, became a self-fulfilling prophecy of relentless advancement.

As computers became personal in the '80s and '90s, they migrated from exclusive research labs to offices, homes, and ultimately our pockets. Software flourished, networks sprawled, and the World Wide Web interconnected humanity in an intricate digital tapestry. Each innovation harnessed the ever-growing power of hardware, creating a feedback loop of mutual advancement.

With the dawning of the new millennium, data became the new oil. Pervasive computing devices, from smartphones to sensors, churned out vast volumes of information. The challenge shifted from mere computation to making sense of this data deluge. Thus, the seeds for the rise of artificial intelligence were sown.

Machine learning, a subset of AI, began leveraging sophisticated algorithms to discern patterns from data. These patterns, too complex for human cognition, revealed insights, powered recommendation engines, and fueled virtual assistants. Yet as powerful as these algorithms were, they were but precursors to the deeper magic of neural networks.

Inspired by the human brain, deep learning neural networks began replicating its intricate web of neurons and synapses, albeit in a simplified manner. Suddenly, machines weren't just processing; they were learning, recognizing, and, in some sense, thinking. Tasks once believed to be in the exclusive domain of human cognition—like image and speech recognition—became playgrounds for these algorithms.

Today, as we stand on the precipice of the AI revolution, we marvel at how far we've come—from punch cards that merely counted to neural networks that dream. The odyssey of computer power, it seems, is an ever-unfolding tale, with horizons that keep expanding into the vast expanse of possibility.

A Connected World

Now more than ever, we live in a connected world, otherwise referred to as the "IoT," or Internet of Things.

According to IBM, "The Internet of Things (IoT) refers to a network of physical devices, vehicles, appliances and other physical objects that are embedded with sensors, software and network connectivity that allows them to collect and share data."[30] This connectivity allows these devices to interact with one another, with central systems, and with users, often in real time.

These are physical devices embedded with sensors, software, and other technologies to connect and exchange data with other devices or systems. Examples include smart thermostats, wearable health devices, connected cars, and even smart refrigerators.

IoT devices continuously gather data from their environments. This data is then sent to centralized systems, cloud platforms, or other devices for analysis. By processing the data, these devices can make semiautonomous decisions. For example, a smart thermostat can learn a user's preferences over time and adjust heating or cooling accordingly without direct user input.

> *Most task-based communications have focused on 1-on-1 type engagements, and it is clear that future advancements in the field of AI will emphasize more multi-speaker team-based communications, especially in working to achieve solutions that leverage experts from a diverse range of areas.*
>
> ♀ **DR. JOHN HANSEN,**
> Associate dean for research, professor of electrical engineering and speech sciences, University of Texas, Dallas

IoT enables richer interactions between devices and users. For instance, a wearable health monitor can provide feedback on a person's physical activity, sleep patterns, and heart rate, helping them make informed health decisions.

Everything Is Connected

The phrase "Everything is connected" captures the essence of a world where virtually every device, system, and possibly even living organisms are interconnected in some manner. It's an extension of the IoT concept but on a more comprehensive scale.

As more devices get connected, they often rely on each other for better functionality. For example, a smart home may have interconnected lights, security systems, thermostats, and appliances, all working together to provide an optimized living experience. The connectivity ensures that data can be shared, almost instantly, between devices. This allows for real-time monitoring and adjustments. For instance, a smart grid can adapt to changes in energy demand dynamically.

Beyond individual devices, "Everything is connected" also points to broader systems like transportation networks, power grids, and cities becoming integrated, smart, and responsive. The convergence of IoT and the concept of "Everything is connected" is transforming industries, reshaping how we live, and redefining the boundaries between the physical and digital worlds. While this interconnectivity brings immense opportunities for efficiency and innovation, it also presents challenges related to security, privacy, and the management of vast amounts of data.

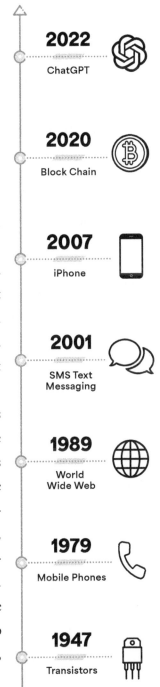

2022
ChatGPT

2020
Block Chain

2007
iPhone

2001
SMS Text
Messaging

1989
World
Wide Web

1979
Mobile Phones

1947
Transistors

Conclusion

With the rise of digitization, the rules of the game have become ever more complex.

In their *Harvard Business Review* article, Paul Leinwand and Mahadeva Matt Mani write, "Despite so much energy and investment in digitization, we are hearing many executives express concern that they are actually falling behind on making the important choices that lead to differentiation. They're right to worry, because winning in the post-Covid world will require re-imagining not just how you work, but also what you do to create value in the digital era."[31]

However, to understand how it's possible to win in this digital age, we need to pause and understand the fundamentals of artificial intelligence.

CHAPTER 3:

FUNDAMENTALS OF MODERN ARTIFICIAL INTELLIGENCE

ARTIFICIAL INTELLIGENCE (AI) REFERS TO A BRANCH OF computer science dedicated to creating systems capable of performing tasks that typically require human intelligence.

In the words of John McCarthy, AI "is the science and engineering of making intelligent machines, especially intelligent computer programs. It is related to the similar task of using computers to understand human intelligence, but AI does not have to confine itself to methods that are biologically observable."[32]

As we stand on the cusp of technological advancements that seemed like science fiction just a few decades ago, it is unavoidable for professionals across all sectors to understand the principles of AI and engage with the crucial conversations that will shape the future of this technology and, by extension, the world.

Most modern AI systems fall under one of two categories: Narrow AI and General AI. Narrow AI, or Weak AI, systems are designed for specific tasks and are limited in terms of versatility. Examples include voice assistants, image recognition programs, and recommendation engines. We might think of these features as *assistants*.

General AI, or Strong AI, can successfully perform any intellectual task that a human being can. It is a flexible form of intelligence capable of learning how to carry out vastly different tasks, adapt to new circumstances, and understand the world around it. General AI isn't just an assistant to humans; in many cases, it is a *replacement*. As of now, this type of AI is theoretical and does not yet exist.

We've seen many significant AI advancements in recent years, and as we delve into the core of modern AI, it is paramount to first understand its fundamentals. The landscape of AI has dramatically evolved, revolutionizing industries, economies, and societal structures. In this chapter, we will unpack the basic yet integral components that constitute modern AI, explore various types of AI, and understand the underpinning methodologies that enable machines to mimic cognitive functions that are synonymous with the human mind, such as learning and problem-solving.

Deep Learning Algorithms and Generative AI

One of the major breakthroughs in modern AI is the development of deep learning algorithms. These algorithms, inspired by the structure and function of the human brain, enable machines to learn from vast amounts of data and make intelligent decisions. Deep learning has led to remarkable achievements in areas such as image recognition, natural language processing, and speech recognition.

Several methodologies drive the functionality of AI, with the following being the most pivotal in modern AI systems:

Machine learning: At the heart of modern AI, machine learning enables systems to learn and improve from experience without being explicitly programmed. It utilizes statistical techniques to give computers the ability to "learn" through exposure to data.

Neural networks: Inspired by the human brain's architecture, these networks consist of interconnected layers of algorithms, known as neurons, that feed data into each other and that can be trained to carry out specific tasks by modifying the importance attributed to input data as it passes between the layers.

Deep learning: A subset of machine learning, deep learning allows computational models that are composed of multiple processing layers (hence "deep") to learn representations of data with multiple levels of abstraction. It is crucial for tasks such as image and speech recognition.

Natural language processing (NLP): This AI methodology focuses on the interaction between computers and humans through language. It allows machines to read text, hear speech, interpret it, measure sentiment, and determine which parts are important.

Computer vision: This field of AI trains computers to interpret and understand the visual world. Using digital images from cameras and videos and deep learning models, machines can accurately identify and classify objects and then react to what they "see."

Another important aspect of modern AI is the rise of generative AI: "Generative AI builds on existing technologies, like large language models (LLMs) which are trained on large amounts of text and learn to predict the next word in a sentence."[33] Generative AI models, such as GPT-4, can generate humanlike text, music, and art. These models have opened new possibilities in content creation, creative industries, and even virtual assistants.

Generative AI models are a part of deep learning and are typically built using foundation models, which are expansive artificial neural networks inspired by the human brain. These models can perform several functions, such as classifying, editing, summarizing, answering questions, and drafting new content. They can create novel data based on existing patterns, which allows them to excel in a wide range of applications including natural language processing, image synthesis, and more.

Some of the key generative AI models include Restricted Boltzmann Machines (RBMs), Variational Autoencoders (VAEs), Generative Adversarial Networks (GANs), Recurrent Neural Networks (RNNs), Long Short-Term Memory (LSTMs), and Transformers.

Generative AI models can generate new data based on the patterns and distributions observed in the training data. They are used in a wide range of applications and have demonstrated significant potential in creating high-quality content in minutes, which would otherwise take days or weeks for humans to produce. However, they also present challenges, such as the computational power required for training and the need for increased responsibility in AI usage.

I should add that AI has also made significant strides in the field of robotics. Advanced robots equipped with AI algorithms can perform complex tasks, such as autonomous navigation, object manipulation, and even humanlike interactions. This has implications for industries such as manufacturing, healthcare, and transportation.

Machine Learning and Neural Networks

This brings us to a couple of interconnected concepts: machine learning and neural networks.

In her MIT Sloan article, Sara Brown writes, "Machine learning is a subfield of artificial intelligence, which is broadly defined as the capability of a machine to imitate intelligent human behavior."[34] Machine learning turns the traditional programming paradigm on its head. Instead of coding explicit rules for computers to follow, it employs algorithms that learn and make inferences directly from data. These systems improve their performance as the amount of data available for learning increases.

Neural networks are a step forward in the complexity and capability of machine learning models. They consist of layers of nodes, or "neurons," each performing a simple computation and transmitting the result to subsequent layers, a structure inspired by biological neural systems. These networks excel in capturing nonlinear relationships and are particularly potent in handling unstructured data such as images, audio, and text. Their capacity for deep learning, where neural networks have many layers, allows for the tackling of complex problems ranging from speech recognition to medical diagnosis, proving essential in numerous technological advancements.

Convergence and the interdependence between machine learning and neural networks is undeniable, as neural networks represent a kind of machine learning architecture. The convergence occurs in their shared goal: learning from data to make decisions or predictions. Machine learning provides the foundational frameworks and principles, including the algorithms that enable the learning process. Alternatively, neural networks contribute to the layered architecture that allows for more nuanced, humanlike processing and deep learning, fostering advancements in image and speech recognition, among other areas.

Machine learning and neural networks serve as the backbone for countless innovations that once seemed the stuff of science fiction. Their capabilities, continually enhanced by burgeoning data and increasing computational power, hold immense potential to shape the future.

However, these powerful new technologies are not without limitations. Data quality and bias can also impact the performance of machine learning and neural networks. If the training data is biased or of poor quality, the model may learn and perpetuate those biases, leading to unfair or discriminatory outcomes. They also require substantial computational resources and extensive training data to achieve optimal performance. Ultimately, it is also difficult to understand the reasoning behind their predictions, making machine hallucinations and biases all too common.

Furthermore, the limitations of machine learning and neural networks extend to their inability to reason or understand context in the same way humans do. While these models excel at pattern recognition and making predictions based on data, they lack the cognitive abilities and common-sense reasoning that humans possess.

Alongside this optimistic trajectory, it is crucial to navigate the ethical, privacy, and employment landscapes altered by these technologies. By addressing these challenges head-on, society can harness the full potential of machine learning and neural networks in a manner that is both progressive and humane, steering the future toward a horizon of inclusive, ethical, and comprehensive advancement.

Natural Language Processing (NLP) and Natural Language Understanding (NLU)

Natural Language Processing (NLP) and Natural Language Understanding (NLU) are two related, but distinct, concepts in the field of artificial intelligence.

They are crucial components of artificial intelligence that enable computers to comprehend and interpret human language. NLP and NLU are interdependent, with NLU building on the foundations laid by NLP. NLP deciphers the linguistic structure, while NLU extracts context—a collaboration that allows for sophisticated dialogue systems and virtual assistants capable of interpreting human language with an unprecedented degree of clarity.

"NLP drives computer programs that translate text from one language to another, respond to spoken commands, and summarize large volumes of text rapidly—even in real time."[35] It involves techniques such as sentiment analysis, text classification, and language translation. NLP focuses on analyzing language at various levels, from individual words and phrases to entire sentences and paragraphs. It aims to address the complexity and uncertainty of natural language by using machine learning algorithms, such as deep neural networks and decision trees, to recognize patterns in text and make predictions or perform tasks.

The importance of NLP in customer service, for example, cannot be overstated. It allows for the analysis of consumer interactions, resulting in increased customer satisfaction and efficiency. For example, voice intelligence technology, powered by NLP, can accurately transcribe calls in real time and track keywords and their frequency in consumer interactions.

Natural Language Understanding (NLU) specifically focuses on the task of understanding the meaning and context of human language. It involves the ability of computers to comprehend and interpret the intent behind a piece of text or speech. NLU enables computers to extract relevant information, identify entities, and understand the relationships between different parts of a sentence or conversation. It focuses on enabling computers to understand the meaning and context of human language and deals with the nuances of human language.

Applications that require chatbots to understand conversational language and respond appropriately include the following:

- **Virtual assistants:** The rise of virtual assistants like Siri, Alexa, and Google Assistant epitomizes NLP's and NLU's success, capable of processing and responding to voice commands and inquiries.

- **Customer service:** AI-driven chatbots, equipped with NLP and NLU, handle customer inquiries, complaints, and other interactions, providing timely and contextually relevant responses.

- **Healthcare:** From processing patient records to facilitating more natural human-machine interaction in mental health therapy, these technologies are enhancing healthcare provision.

- **Language translation services:** NLP and NLU are crucial in real-time speech translation and text translation services, enabling clearer, context-aware communication across different languages.

As with all AI, NLP and NLU have limitations, such as the complexity of human language itself. With multiple ways to express the same idea and various meanings for the same word depending on the context, NLP models may struggle with understanding and interpreting these nuances accurately.

NLP and NLU play a vital role in enabling computers to understand and interpret human language. They have numerous applications, particularly in customer service, where they enhance efficiency and customer satisfaction. However, it is important to be aware of the limitations of NLP, including the complexity of language, the need for extensive training data, and potential biases. By understanding these limitations, we can develop more robust and reliable NLP solutions.

Robotic Process Automation (RPA): Transforming the Intelligent Workforce

This brings us to robotic process automation (RPA).

Director of Product at Nintex Aaron Bultman says, "RPA is a form of business process automation that allows anyone to define a set of instructions for a robot or 'bot' to perform."[36] Robotic process automation (RPA) stands at the forefront of technology-driven operational excellence, embodying the shift toward automation that encapsulates the fourth Industrial Revolution.

So let's explore RPA's nuances, explaining its applications, benefits, and challenges for enhancing productivity across all industries. The key takeaway is the need for strategic integration and responsible implementation practices in harnessing RPA's full potential.

In the digital era, efficiency and productivity have become central tenets for competitive businesses. RPA is a revolutionary technology that has emerged as a key enabler of business performance and agility. It allows organizations to automate repetitive, rule-based tasks, mimicking human interactions with digital systems. It also differs from traditional automation in that it can mimic human actions and integrate seamlessly with existing systems without the need for extensive programming knowledge.

RPA is anchored on the principle of software robots or "bots" executing routine business processes by interacting with applications in the same way that a human would, only significantly faster and without fatigue. These bots capture data, run applications, trigger responses, and communicate with other systems to perform a variety of tasks.

There are basically three types of bots:

- **Attended bots:** These operate on the same workstation as a human, assisting with tasks as needed, and are triggered by specific events or commands.

- **Unattended bots:** These operate autonomously without human intervention, usually on a scheduled basis, and can be used for end-to-end automation.

- **Hybrid bots:** Combining the capabilities of attended and unattended bots, these provide a comprehensive solution, ensuring both efficiency and flexibility.

An RPA software robot never sleeps, makes minimal mistakes, and costs a lot less than an employee. And its versatility extends across various sectors, enhancing operational efficiencies and data accuracy. Organizations use RPA for operations such as accounts payable and receivable, expense management, and fraud detection. It's used in the back office to complete tasks such as data entry, document scanning, and information transfers from one system to another.

RPA software also helps automate tasks like fraud detection, compliance reporting, and customer service requests. It streamlines patient scheduling, billing, claims processing, and report automation, and it enhances supply chain management and customer support and ensures competitive pricing strategies. It improves billing, data management, and network issue resolutions.

RPA tools have strong technical similarities to graphical user interface (GUI) testing tools. These tools also automate interactions with the GUI and often do so by repeating a set of demonstration actions performed by a user. RPA tools differ from such systems that allow data to be handled in and between multiple applications—for instance, receiving email containing an invoice, extracting the data, and then typing that into a bookkeeping system.

The strategic benefits of RPA extend far beyond simple automation, creating opportunities for process improvement, greater operational flexibility, and strategic reallocation of resources toward areas that create true business value. While RPA is a form of digital transformation, it's often seen as a Band-Aid or temporary solution. In the long term, businesses may look to deeper, more integrated solutions with true machine learning and artificial intelligence.

Benefits of Robotic Process Automation (RPA)

RPA offers several strategic benefits that go beyond simple cost cutting. By relieving employees of mundane and repetitive tasks, RPA allows companies to refocus their human resources on tasks that add more value. Below are several strategic benefits:

- ♀ **Accuracy.** RPA reduces the likelihood of errors that might be introduced by human workers, especially in highly repetitive tasks.

- ♀ **Business agility.** Implementing RPA can increase a business's agility, allowing it to better adapt to changing market conditions. With RPA, businesses can easily scale operations up or down and deploy resources where they're most needed, quickly responding to new opportunities or challenges.

- ⚙ **Cost efficiency / savings.** By automating repetitive tasks, businesses can often achieve significant cost savings, as bots can work twenty-four seven without the need for breaks or sleep.

- ⚙ **Customer satisfaction.** By automating routine tasks, services become more consistent and timelier. Faster response times and minimal error processing contribute to improved customer experiences and satisfaction.

- ⚙ **Data quality and analytics.** Automation ensures that data is captured consistently and accurately. Better data quality enables more accurate analytics, which can lead to better strategic decision-making.

- ⚙ **Digital transformation and innovation.** RPA often serves as a stepping stone for more significant digital transformation within a company, acting as a catalyst for adopting newer, more advanced technologies and innovations. It creates a culture that embraces automation and encourages continuous process improvement.

- ⚙ **Employee satisfaction and retention.** Removing monotonous tasks from employees' workloads enables them to focus on more engaging, creative, and higher-value tasks, leading to increased job satisfaction and retention.

- ⚙ **Enhanced compliance.** RPA can be programmed to meticulously follow all the rules set by a company, ensuring processes are executed in a compliant manner. This level of consistency is particularly important in regulated industries.

- ⚙ **Increased productivity.** RPA operates twenty-four seven without interruptions, significantly reducing the time required to complete tasks. This continuous workflow can substantially increase productivity compared to human workers performing the same tasks.

- ♀ **Integrations.** RPA can interact with various applications in the same way a human user might–by logging in, entering data, clicking buttons, reading data, etc. This makes it possible to integrate systems without the need for traditional API-based integration.

- ♀ **No code / low code.** One of the hallmarks of many RPA tools is that they do not require deep coding or programming skills. Instead, they provide a graphical interface where users can design automation workflows.

- ♀ **Scalability.** Organizations can deploy multiple bots to handle high-volume tasks, making it scalable.

In summary, the strategic benefits of RPA extend far beyond simple automation, creating opportunities for process improvement, greater operational flexibility, and strategic reallocation of resources toward areas that create true business value.

Conclusion

In 2018, Janna Anderson and Lee Raine wrote a Pew Research article and said that in the future, "AI will be integrated into most aspects of life, producing new efficiencies and enhancing human capacities."[37] As the years have progressed, we've seen this statement become reality.

It's easy to bury our heads in the sand and ignore what's ahead. But a better approach is to understand what we're dealing with and develop a game plan for our organizations to leverage modern AI to our advantage.

> In a future of human-AI collaboration, everyone will need a basic grasp of AI literacy and data analysis. The ability to adapt and learn continuously will be vital in the face of evolving tech-

nology. Skills that uniquely amplify the human role in this new human and machine world include critical thinking, creativity, emotional intelligence, and strong communication. These will be essential to problem-solving, innovation, and ensuring ethical use of AI. Hybrid skills with a technological slant will be in high demand in the future.

💡 MIKE MATEER,
Senior vice president, IT operations and services, Maximus

To do this, we need to address the relationship between humans and machines.

CHAPTER 4:

HUMANS AND MACHINES–THE COLLABORATION PARADIGM

WHEN IT COMES TO THE DISCUSSION SURROUNDING HUMANS and machines, the one word you often don't hear is *collaboration*. Instead, many people fall into one of two categories: either they are *all in* and want to use AI for everything, or they are *all out* and would love to see a world without it. They don't see how humans can work with machines to make a powerful impact.

According to a 2022 Pew Research poll on how Americans think about artificial intelligence, 37 percent of Americans said they were more concerned than excited, 18 percent were more excited than concerned, and 45 percent were equally concerned and excited.[38] Skepticism is high, but according to *Forbes* contributor Anthony

Lancaster, "Humans *shouldn't* be afraid of algorithms taking over the world. Instead, the mindset should be collaborative—a means for helping its users improve at their jobs and continue doing what they already do best."[39]

Humans and machines can work together and form a symbiotic relationship in a healthy ecosystem by capitalizing on each other's strengths. Machines excel at handling repetitive tasks, analyzing data, and processing vast amounts of information at high speeds, while humans bring creativity, emotional intelligence, and complex problem-solving skills. Striking a balance in this partnership involves effective communication, a clear delineation of roles, and ongoing adaptation to ensure a harmonious coexistence in the evolving landscape of work and technology.

Strengths and Weaknesses of Machines and AI

First, it's important to acknowledge the pros and cons of machines and artificial intelligence. Machine learning and AI bring a host of strengths and some weaknesses to the table. Let's start with the strengths.

AI can process and analyze data much faster than a human can, enabling quick decision-making when dealing with large datasets. Likewise, machines can operate continuously without the need for breaks, sleep, or vacations, leading to significant improvements in productivity. AI systems can also be scaled up to handle an increasing volume of work without a corresponding increase in errors or fatigue.

For tasks involving pattern recognition or data processing, AI can work with high accuracy and consistency, often surpassing human performance. Over time, AI can be more cost-effective than human labor, as it does not incur the same recurring costs (like salaries and benefits). AI systems are adept at managing vast amounts of information, sorting through it, and extracting relevant insights. Machines can perform tasks without the cognitive biases that humans may have, potentially leading to more objective decision-making. They can handle dangerous tasks that would be unsafe for humans, such as working in toxic environments or handling heavy machinery.

AI and machines are exceptionally efficient. They excel in pattern recognition and are cost-effective over time. They are especially suited for hazardous tasks where human safety is a concern.

But there are also weaknesses. While AI can perform specific tasks with remarkable precision, it cannot fully replicate human intelligence and creativity. AI lacks consciousness and emotions, limiting its ability to understand complex human experiences and produce truly creative works. It can only generate output based on its programming and the data it has been trained on. Most AI systems are designed for specific tasks (narrow AI) and struggle with generalizing their knowledge to new, unanticipated scenarios. AI does not possess emotional intelligence and cannot make ethical decisions or understand human emotions in a meaningful way.

AI's performance is heavily dependent on the quality of the data it is fed, and poor-quality or biased data can lead to inaccurate or unfair outcomes. The initial setup, development, and training of AI can be expensive and resource-intensive. Also, while AI can learn from new data, it generally lacks the human ability to adapt quickly to entirely new contexts or changes in the environment.

Another downside is that many AI models, especially deep learning algorithms, are often seen as "black boxes" with decision-making processes that are not transparent or easily understood. AI systems also require regular, complex, and costly updates. They can displace human jobs and can be vulnerable to data breaches, which pose massive privacy concerns. They lack the creative and intuitive problem-solving abilities inherent to humans and are heavily reliant on the quality of input data, which can lead to inaccuracies or biases. Their flexibility is limited, and they often struggle with tasks outside their specific programming.

Also, initial setup and maintenance costs for AI and machines are high. Additionally, the risk of job displacement in certain sectors and the environmental impacts of these technologies are concerning. While AI and machines are powerful tools for data handling and operational efficiency, their limitations in creativity, ethical complexities, and societal impact must be carefully managed.

Why Augmentation Is Better Than Replacement

In our pursuit of technological advancement, it is imperative to recognize that machines are not a wholesale replacement for human capability. Rather, we should embrace the concept of augmentation, a strategic collaboration between humans and machines, often referred to as "augmented intelligence." This paradigm harnesses the unique strengths of both entities to yield optimal outcomes.

Within the paradigm of augmented intelligence, we have the Human in the Loop (HITL) approach, which stands out as a pivotal methodology. Here is how this works: human intervention becomes indispensable when machines encounter challenges, creating a dynamic feedback loop that fosters continuous AI improvement. This involvement extends to tasks such as result validation, cleaning, and correction, where human expertise proves invaluable.

The integration of HITL is not merely a technical nuance but also a fundamental principle shaping the future of work in the era of artificial intelligence. The synergy of humans and machines *amplifying* AI capabilities, rather than *supplanting* them, serves as our guiding ethos. Humans enhance AI systems in many ways.

First, while AI excels at optimization *within* defined parameters, human ingenuity shines in thinking *outside* the box to devise creative solutions for intricate problems. Humans strategically utilize AI for routine tasks, liberating cognitive resources for innovative and creative pursuits.

Moreover, the role of humans extends to the training of AI systems, providing nuanced contextual data that forms the foundation for machine learning algorithms. Beyond initial training, humans persist in refining AI by offering feedback and rectifying errors. In areas demanding empathy, such as healthcare or customer service, humans integrate AI for data processing while ensuring a crucial human touch in interactions.

Decision-making underscores another facet of human-AI collaboration. AI offers a spectrum of actions based on data analysis, yet the final decisions require human interpretation, factoring in elements beyond AI's purview, such as long-term strategic goals and the intricacies of human relationships. Humans also bring a more personalized approach. They help tailor services and products based on a deep understanding of individual needs and preferences—something AI alone cannot fully grasp.

> Smart companies will understand that AI and automation aren't going to replace humans anytime soon. Many companies made an error in judgement when chatbots came onto the scene, thinking they were the end of the Customer Service department. We can't make that mistake again. AI must continue to work with humans, integrating the best of what both have to offer. This will result in the elimination of repetitive tasks, ease of

information retrieval, and the ability of humans to do what they do best—be human.

DAN GINGISS,
Chief experience officer, The Experience Maker

Acknowledging AI systems' maintenance and adaptability demands, humans assume the vital role of overseers, ensuring correct and efficient functioning. Human intuition and experience become paramount for guiding AI systems through adaptation and evolution in situations devoid of historical data. In the linguistic domain, humans bring an unparalleled understanding of nuance, sarcasm, humor, and subtlety. In language-related tasks, humans guarantee communication clarity, appropriateness, and effectiveness, enriching the interaction in ways that AI struggles to emulate.

Having humans in the loop is indispensable for training, managing, and ensuring the responsible use of AI. This partnership paves the way for AI systems that perform better and align more closely with human needs and societal values.

Human Intuition and Machine Precision

When discussing augmentation, we're getting the best of both worlds. We're valuing human *intuition* and machine *precision*.

Human intuition and machine precision represent two very different yet complementary facets of problem-solving and decision-making. In human-machine collaboration, human intuition can guide and complement machine precision, ensuring data-driven and human-centered decisions. This synergy allows for nuanced and sophisticated outcomes that neither could achieve alone.

Human intuition contributes a nuanced understanding of context, ethical considerations, and the subtleties of human behavior and society. It allows for flexible decision-making and the ability to perceive abstract concepts and relationships, something AI alone might miss. Humans can also provide creative problem-solving that goes beyond the data, envisioning new ways to apply AI or identifying novel solutions to complex problems.

On the other side, machine precision offers the ability to handle vast quantities of data with speed and accuracy unmatched by human capabilities. It brings consistency to operations, eliminating the errors and biases that human operators may introduce. AI excels in recognizing patterns, optimizing processes, and delivering quantifiable insights that can inform better decision-making.

When these two forces come together, AI is dramatically improved. Human intuition guides machine learning models toward more relevant and context-aware applications, ensuring that the data is interpreted correctly and the AI's recommendations are applicable in the real world. Meanwhile, machine precision provides humans with the tools to make informed decisions quickly and effectively, supporting intuition with hard data and clear patterns.

In practice, this collaboration can result in more efficient AI aligned with human values and needs. It leads to AI systems that are trusted and integrated seamlessly into daily operations, augmenting human abilities and allowing for an expansion of what can be achieved through technology.

The synergy between human intuition and machine precision creates a powerful dynamic. They complement each other in ways that significantly enhance the capabilities of AI. Combining human intuition and machine precision creates a balanced AI approach that maximizes efficiency, fosters innovation, and maintains ethical standards. This partnership improves AI and enables it to be used more effectively and responsibly across various domains.

UNDERSTANDING THE DIFFERENCES BETWEEN INTUITION AND PRECISION

HUMAN INTUITION

- **Qualitative**: Intuition is often based on qualitative insights rather than hard data.

- **Experiential**: It's derived from personal experiences and subconscious information processing.

- **Adaptive**: Intuition allows for quick adaptation to new situations without explicit reasoning.

- **Contextual**: Humans can pick up on social cues and contexts that may be invisible to machines.

- **Innovative**: Intuitive leaps can lead to innovative ideas and solutions beyond the obvious.

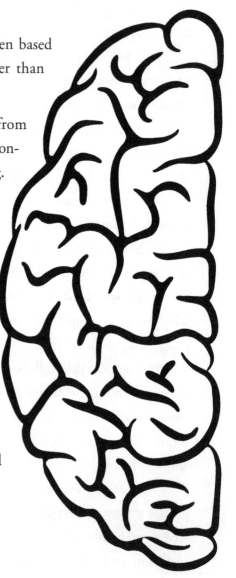

MACHINE PRECISION

•**Quantitative**: Machines operate with precision, using quantitative data to make decisions.

• **Consistency**: They perform tasks consistently and can repeat the same processes without variation.

• **Speed:** Machines process information and perform tasks at speeds unattainable by humans.

• **Scalability**: Precision and speed can be scaled up, handling vast amounts of data and operations.

•**Reliability**: Machines are reliable in the quality of the work, given that the input data and algorithms are correct.

THE POWER OF COLLABORATION

In our collaborative journey between humans and machines, human intuition emerges as a crucial element, contributing emotional intelligence, ethical understanding, and real-life sentiment analysis.

Humans excel at interpreting decisions, fine-tuning machines for editorial or ethical content, and infusing a "human touch" that machines currently lack the finesse to replicate. This human input serves as a guiding force, steering machines toward more effective computations of real-life scenarios over time.

Another integral aspect of this collaboration lies in the evolution of Collaborative AI Systems. These systems seamlessly operate alongside humans in real time, engaging in tasks like cowriting or codesigning. They capitalize on the strengths of both human and AI models, generating content with enhanced effectiveness, efficiency, and creativity.

Then there is Interactive Machine Learning. This entails designing AI systems that seamlessly integrate human interaction, preference, and judgment, aiming to craft more useful and meaningful AI systems that augment human capabilities. The potent synergy between human intuition and machine precision acts as a catalyst for enhancing the overall capabilities of AI systems.

It's important to note that AI holds the potential to augment human capabilities, streamlining tasks for increased efficiency. For instance, in medicine, AI assists doctors in diagnosing diseases with greater accuracy. Conversely, AI replacing humans can lead to job obsolescence, sparking unemployment and economic disruption concerns. Yet with job elimination comes the potential for innovation and increased efficiency, opening new opportunities.

There is also room for collaboration in skill development and education. Augmenting humans with AI demands continuous learning and adaptation to new technologies. This collaborative approach fosters a culture of lifelong learning but also necessitates significant changes in education systems.

> *In the medical device industry, the collaborative partnership between human expertise and machine intelligence cultivates an advanced workforce, driving the creation of innovative solutions that elevate patient care, improve outcomes, and set new standards in healthcare technology.*

> ♀ **DOUG ELLISON,**
> Chief revenue officer, Micro Transponder

When it comes to ethical considerations, when AI *augments* human capabilities, questions arise about responsibility, privacy, and control. For example, who is responsible for decisions made with AI assistance? When AI *replaces* humans, ethical concerns shift to broader societal impacts, such as increased inequality and the potential loss of certain skills and crafts.

Examining economic impacts, AI augmentation can stimulate new job creation and economic growth in sectors where human-AI collaboration is essential. Conversely, AI replacing humans in certain jobs may have negative economic consequences, especially for lower-skilled workers. In terms of social and psychological effects, augmenting humans with AI can reshape the nature of work and social interactions, potentially leading to challenges like increased dependence on technology and shifts in social dynamics. Replacement, however, may have more profound effects, including the erosion of certain social structures and community roles.

Regulatory and governance challenges loom large for both augmentation and replacement by AI, requiring robust frameworks to ensure safety, privacy, fairness, and accountability. The specifics of these frameworks might vary significantly depending on whether AI is augmenting or replacing human roles. Ultimately, the balance between AI augmentation and replacement hinges on carefully considering these factors, aiming to maximize benefits while minimizing potential harms.

The Future Is Collaborative

The myth that AI can or will completely replace human workers is widespread, but AI is better suited to augment human capabilities rather than replace them.

Ultimately, the relationship between humans and AI is not a zero-sum game; it's synergistic. Humans augment AI by providing the qualities that AI lacks, such as creativity, empathy, ethical reasoning, and contextual understanding. Meanwhile, AI augments human capabilities by handling large-scale data analysis, performing repetitive tasks, and extending our cognitive and analytical capacities. This collaboration leads to a more productive and innovative future, with AI acting as a tool that empowers human potential.

> In the future of business, AI and Humans will collaborate not as competitors but as partners, leveraging the strengths of each to drive innovation, enhance decision-making, and optimize performance, ultimately forging a path toward unparalleled success and sustainable growth.
>
> ♀ **SCOTT HERMANN,** Founder and CEO, IDIQ

The collaboration between humans and machines is a promising paradigm that can lead to success in a variety of contexts, sectors, and geographic settings. The key to success in human-machine collaboration within AI is to balance the scale between technological capabilities and human virtues, creating an environment where each complements the other, leading to a sum greater than its parts.

CHAPTER 5:

THE ROLE OF HUMANS IN AN INTELLIGENT WORKFORCE

THE ROLE OF HUMANS IN AI AND THE INTELLIGENT WORKFORCE is multifaceted and critical to the success and effectiveness of AI systems.

Humans are essential for providing the emotional intelligence, ethical understanding, intent, and conversational design that AI and machines lack. Human workers are expected to focus on innovating, adding value, and pursuing new skills, complementing the capabilities of AI and machines. While some see AI as a replacement for humans, this is an unlikely proposition. In their *Harvard Business Review* article, David De Cremer and Garry Kasparov write the following:

> The question of whether AI will replace human workers assumes that AI and humans have the same qualities and abilities—but, in reality, they don't. AI-based machines are fast, more accurate, and consistently rational, but they aren't intuitive, emotional, or culturally sensitive. And, it's exactly these abilities that humans possess and which make us effective.[40]

The intelligent workforce is about leveraging the strengths of both humans and technology to drive business growth and adapt to future changes. Humans are also responsible for making creative decisions and interpreting data-driven insights to inform business strategies. As technology evolves, human roles may change, but their value increases, especially in areas where nuanced understanding and emotional intelligence are required.

The intelligent workforce aims to augment human performance with digital technology, ensuring that humans and machines work together to achieve the best outcomes. This collaboration is a way to liberate human potential and focus on high-value tasks that machines cannot perform.

AI has "learned" to generate artwork based on human-created artwork, but what would take a human artist countless hours to create, AI can generate in a matter of seconds

Practical Ways to Keep "Humans in the Loop"

There are many ways to ensure that human intuition remains a priority and humans actively participate in the decision-making or operational processes alongside automated systems or artificial intelligence.

One of these is design and development. Humans are essential in designing and developing AI systems. They define the problems AI needs to solve, design algorithms, choose appropriate datasets for training, and develop the overall architecture of the AI system. This process requires not only technical expertise but also an understanding of the ethical, social, and practical implications of AI applications.

Another is training and supervision. AI systems, especially those based on machine learning, require training with large datasets. Humans are involved in curating and labeling these datasets, a process crucial for the accuracy and reliability of AI predictions and decisions. Post-development, human oversight is necessary to supervise AI operations, ensuring that they function as intended and adhere to ethical standards.

Then there are the all-important factors, such as interpretation and decision-making. While AI can analyze vast amounts of data and identify patterns beyond human capability, humans are needed to interpret these findings within the context of real-world applications. Critical decision-making, especially in complex or ethically ambiguous situations, often requires human judgment complemented by AI insights.

It's also important to note that humans play a vital role in establishing ethical guidelines for AI use and ensuring compliance with legal and regulatory standards. This includes addressing concerns like data privacy, bias in AI algorithms, and the broader societal impacts of AI technology.

As AI systems evolve, human workers need to adapt and learn new skills to work effectively alongside AI. This includes understanding AI capabilities and limitations, learning to use new tools and interfaces, and adapting to changes in job roles and industry practices. AI is not infallible and can make errors, especially in situations deviating from training data or complex contexts requiring nuanced understanding. Human intervention is necessary to identify, correct, and learn from these errors, ensuring continuous improvement of AI systems.

AI can assist in providing personalized experiences in customer experience and customer-facing roles, but human empathy and understanding are irreplaceable for building trust and handling sensitive or intricate customer needs.

This brings us back to the importance of collaboration. In many sectors, the optimal approach is collaborative, where AI and humans work together to enhance performance. AI can handle large-scale, repetitive tasks, while humans focus on creative, strategic, or complex problem-solving aspects. This synergy can lead to improved efficiency and innovation.

Emotional Intelligence and Decision-Making

Human emotional intelligence and decision-making significantly impact AI's design, development, deployment, and interaction in several ways.

Emotional intelligence, as defined by Brandon Goleman, "is the capacity to be cognizant and in control of one's emotions and to express these emotions appropriately."[41] It allows humans to provide the nuanced understanding and ethical oversight that AI lacks. This human input is critical for tuning AI systems to handle real-life scenarios with the appropriate emotional and ethical considerations. For instance, AI can analyze tone and sentiment in customer service, but it requires human emotional intelligence to provide genuinely empathetic responses.

Moreover, human decision-making is essential in guiding AI development and application. Humans decide how AI systems should be designed, what values they should uphold, and when they should be deployed. This includes making judgments about the trustworthiness of AI systems, especially as they become opaquer with the rise of complex models like deep neural networks.

Incorporating human judgment and preferences ensures that AI systems are aligned with human values and can operate effectively in real-world contexts. This human-centered approach to AI design is crucial for creating systems that augment human capabilities rather than replace them, fostering a symbiotic relationship in which humans and AI contribute their unique strengths to achieve the best outcomes.

This collaboration is vital for complex decisions affecting human lives, such as those in medicine, law, or defense, where the consequences of errors and machine hallucinations can be significant. According to the former executive vice president of IBM, John Kelly III, "Man and machine working together always beat or make a better decision than a man or a machine independently."[42]

> *The future of AI and human collaboration promises to redefine productivity and creativity, as integrated teams leverage the unique strengths of both to solve complex problems and innovate at unprecedented speeds.*
>
> ♀ **MATT PIERCE,**
> Founder and CEO, Immediate

In applications where AI is used to interpret or respond to human emotions (like customer service chatbots or mental health assessment tools), the emotional intelligence of the designers and users plays a significant role in ensuring that these systems are sensitive, accurate, and appropriate. Human developers and designers who are emotionally intelligent are more likely to consider the ethical implications and potential biases of AI technologies. They tend to design systems that are more user-friendly, empathetic, and sensitive to diverse user needs and cultural contexts.

Design, Oversight, and Strategy

Human influence is integral to the design of AI systems, as it shapes how these systems interact with the world and with users. The design of AI systems is not just about the development of algorithms or machine learning models; it also encompasses the interactions and behaviors that form the human experience around these models. Human-centered AI design involves considering a wide range of factors, including social, political, ethical, cultural, and environmental aspects that are not typically associated with AI development.

Humans influence AI design by incorporating their judgment, preferences, and values into the system. Moreover, human influence is reflected in the iterative design processes that involve user feedback to refine AI systems. This includes creating interfaces that are understandable and usable by humans, and that augment human abilities rather than attempting to replace them. The goal is to create AI systems that are not only technically proficient but also trustworthy, useful, and aligned with the cognitive abilities and workflows of their users.

The decision to trust and adopt AI technologies is often influenced by human emotions and psychological factors. Emotional intelligence helps in understanding and addressing the fears, apprehensions, and expectations people have regarding AI, which is crucial for its wider acceptance and ethical deployment. Human emotional intelligence influences how AI systems are designed, developed, and interacted with, ensuring these systems are more aligned with human values, ethical standards, and societal norms.

The Need for Human Influence

Human influence is the key to successful design of AI systems and datasets, impacting all aspects of AI from conception to deployment. Humans are required for the following:

1. **Defining objectives and use cases.** Humans determine the goals, objectives, and intended applications of AI systems. This involves identifying specific problems or opportunities where AI can be applied, such as enhancing customer service, improving healthcare diagnostics, or optimizing manufacturing processes.

2. **Data collection and preparation.** The choice of data for training AI models is a human decision. This includes what data to collect, how to label it, and how to prepare it for training purposes. The quality, diversity, and representativeness of this data directly impact the performance and bias of the AI system.

3. **Algorithm design and selection.** Humans decide on the algorithms and techniques to be used in AI systems. This involves choosing from various machine learning approaches like neural networks, decision trees, or support vector machines based on the specific requirements of the task.

4. **Ethical considerations and bias mitigation.** Designers and developers of AI systems must consider ethical implications and actively work to mitigate biases. This includes

ensuring fairness, transparency, and accountability in AI systems to prevent discriminatory outcomes.

5. **User experience (UX) design.** The design of user interfaces and interaction models for AI systems, such as digital agents or virtual assistants, is a human-centered process. It requires understanding user needs, preferences, and behaviors to create intuitive and user-friendly experiences.

6. **Testing and validation.** Humans are involved in testing AI systems to ensure they function correctly, meet performance standards, and are safe for deployment. This often involves assessing the system under various conditions and with diverse datasets to check for reliability and bias.

7. **Regulatory compliance and standards.** Designers must ensure that AI systems comply with relevant laws, regulations, and industry standards, which are established by human governance bodies. This includes considerations around privacy, data protection, and industry-specific regulations.

8. **Feedback and iterative improvement.** Human feedback is essential for the iterative improvement of AI systems. Users' experiences, preferences, and criticisms guide developers in refining and enhancing AI models and interfaces.

9. **Training and education.** The way AI systems are introduced and explained to users, and the training provided for their use, significantly impacts their effectiveness and acceptance. Prompt engineering is an example of how the user must be trained to interact with the machine for successful outcomes. This education process is a human-driven aspect of AI design.

10. **Cultural and social context.** Human designers must consider the cultural and social context in which the AI will operate. This includes understanding regional differences, language nuances, and social norms to ensure the AI system is culturally sensitive and appropriate.

Continuous Learning and Adaptability

Humans can be involved in the AI learning process at different levels, providing labeled data or feedback that helps the AI learn and improve over time. This can be done through games or tasks that generate training data as a side effect of human activities.

Disciplined data management can ensure that AI has continuous access to current and relevant data. Regular updates and expanded datasets used for training the AI to reflect new information, trends, and changing environments will help AI to learn from the latest data and remain effective over time. Reinforcement learning algorithms enable AI systems to learn from their actions and the outcomes of those actions, thereby adapting to new situations and optimizing their performance over time.

As we already discussed, humans in the loop play an active role in validating and calibrating AI systems, addressing potential biases or errors, and enhancing performance. Continuously monitoring the performance of AI systems will help humans identify areas for improvement, design issues, or need for data updates. And regular evaluation against key performance indicators (KPIs) and user satisfaction metrics can guide the updating process.

Effective human engagement leads to cross-disciplinary collaboration, which encourages collaboration between data scientists, domain experts, ethicists, and end users to bring diverse perspectives to AI's development and training. This can lead to more robust, adaptable, and ethical AI systems. Collaboration ensures the AI system learns and evolves when combined with continuously assessing and addressing ethical concerns, including potential biases in AI decision-making. This is crucial to safeguard the AI so that it doesn't develop or perpetuate harmful biases or hallucinate.

Additionally, one of the biggest topics discussed anytime AI regulation arises is the importance of transparent AI mechanisms and algorithms. Understanding and striving for transparency in AI mechanisms is imperative, especially in how AI systems learn and make decisions. AI systems should be designed to be explainable, allowing humans to understand and trust the AI's decisions. This involves creating machine learning techniques that produce more explainable models and enable effective human-AI collaboration. This transparency is crucial for trust and accountability, particularly in critical applications.

Likewise, engaging the broader AI community, including open-source projects, to share knowledge, learn from others' experiences, and stay abreast of the latest developments in AI learning and adaptability will help humans guide AI systems toward effective continuous flexibility.

> *As we think about a future where work will be more augmented and even automated by AI and technology, we must keep our focus on the worker—especially those critical front-line workers across all industries that will need to be up-skilled. When we put our people at the center of this work revolution, we will ensure a future ahead that does not lead to greater social and economic inequalities.*
>
> ♀ **DAMIEN HARMON,**
> Senior executive vice president, customer, channel, and enterprise services, Best Buy

This brings us back to the importance of human engagement. Humans play a critical role in the intelligent workforce and are essential in designing, developing, implementing, and overseeing AI systems. Human creativity, emotional intelligence, and decision-making provide contextual understanding, ethical judgment, and creative problem-solving that AI cannot replicate.

The future of AI and the intelligent workforce is not about replacing humans but augmenting human capabilities, fostering a collaborative environment where both can thrive. This collaboration between humans and machines enhances the capabilities of both, ensuring that the intelligent workforce remains adaptable, innovative, and human-centric.

CHAPTER 6:

TRANSFORMING INDUSTRIES WITH INTELLIGENT WORKFORCES

WITH HUMANS PROVIDING INTUITION AND MACHINES providing precision, the integration of human expertise and machine intelligence will continue to transform every industry.

These changes will carry major economic implications. According to a Goldman Sachs article, "As tools using advances in natural language processing work their way into businesses and society, they could drive a 7 percent (or almost $7 trillion) increase in global GDP and lift productivity growth by 1.5 percentage points over a 10-year period."[43]

Because intelligent automation (IA) combines people and machines, this integration significantly impacts jobs and responsibilities across various industries. As AI and automation technologies advance, they take over routine and repetitive tasks, allowing human workers to shift their focus to more complex, creative, and strategic activities that machines cannot perform.

This is a good thing, and it helps workers focus on what matters most. For example, administrative tasks that were once time-consuming can now be automated, freeing up time for strategic thinking and coaching. This new approach can help companies operate more efficiently, with improved performance and cost reduction.

As a result of these combined forces, we're seeing an emphasis on augmenting human capabilities, enhancing productivity, and fostering innovation. Businesses are investing in reskilling and upskilling programs to ensure their workforce can effectively collaborate with AI and automation tools. The goal is to prepare for future changes and demands, ensuring competitiveness and relevance in a rapidly evolving market.

Big-Picture Industry Benefits

While the obvious benefit to human and machine integration is increased productivity, other noteworthy benefits exist. One of these is the value of data-driven decision-making. AI systems can analyze vast amounts of data much faster than humans. This capability enables businesses to make more informed decisions based on real-time data analysis, leading to improved strategies and outcomes.

Then there is customization and personalization. In industries like retail, entertainment, and healthcare, the combination of human insight and machine learning allows for highly personalized products and services. This tailoring to individual preferences and needs enhances customer satisfaction and loyalty. This leads to innovation and new opportunities that open possibilities in product and service innovation.

> *AI is already turning the golf industry upside down, particularly in coaching. I'm most excited about the transformative potential of human-machine collaboration in learning golf. From AI-driven swing analysis, to building a perfect swing based on a person's physical characteristics, the synergy between human expertise and technological innovation promises to revolutionize how people learn to play golf.*
>
> ♀ **AARON BERGMAN,**
> Founder and chief executive officer, Golfinity

As machines take over more routine tasks, the demand for advanced technical skills and soft skills like critical thinking, creativity, and emotional intelligence will rise. This will lead to a significant transformation in workforce skill requirements. But the good news is that in sectors like manufacturing, construction, and mining, machines can perform hazardous tasks, reducing workplace injuries.

Another important factor is that automation and AI can reduce costs by optimizing resource use, reducing waste, and improving process efficiencies. These savings can then be reinvested in other areas of the business. Along with this, AI and automation technologies enable businesses to scale operations quickly in response to market demands. They also provide the flexibility to adapt to changing market conditions or customer needs.

Because intelligent machines can optimize energy use and reduce waste by reducing defects, this contributes to more sustainable business practices and helps industries reduce their environmental footprint. Thus, this intelligent workforce's digital and connected nature facilitates global collaboration, allowing human and machine teams to work together across geographical boundaries.

> The evolution of technology has always played a role in the music-making process, from the advent of the electric guitar, to synthesizers, digital recording and now AI is the most powerful tool yet. Art is a reflection of the human experience and people who will succeed in the future marketplace are the ones who can merge the power of AI with the power of human creativity however possible.

♀ **RYAN CHAVEZ,**
Executive producer, founder, and CEO, Imprinted Group

The intelligent workforce and the synergy between humans and machines transform existing industries and pave the way for new business models and opportunities. Let's explore a few specific industry examples.

EXAMPLE 1: HEALTHCARE– DIAGNOSTICS AND TREATMENT

The first of these is healthcare.

AI and the human workforce are changing healthcare diagnostics and treatments by creating a more integrated and efficient approach to patient care. AI algorithms are increasingly used to read diagnostic scans with high accuracy, assisting doctors in diagnosing patient cases more quickly and identifying suitable treatments. This integration allows healthcare professionals to focus on the more nuanced aspects of patient care, such as developing treatment plans and providing empathetic support to patients.

In addition, AI-driven diagnostics are limited to image analysis and extend to predictive analytics, where AI can help identify patients at high risk for certain conditions, such as high-risk pregnancies. This predictive capability enables earlier interventions, leading to better health outcomes. According to a Johnson & Johnson article, "When it comes to catching and diagnosing diseases earlier, AI can be a real game changer. By applying AI to data derived from or generated by common diagnostic tests, such as electrocardiograms and echocardiograms, providers could diagnose diseases more accurately, prevent delays in care and potentially save lives."[44]

Moreover, AI is being used to streamline administrative tasks like filling out insurance paperwork, which frees healthcare workers to spend more time with their patients. This shift toward higher-value work is crucial in a field where empathy and human interaction are paramount.

Integrating AI into healthcare also facilitates the development of new treatments and medication delivery systems, where the combination of human expertise and advanced algorithms can lead to more personalized and effective care strategies. Healthcare providers are employing AI to scan x-ray and CT images to diagnose conditions such as pneumonia. This integration allows for quicker and more accurate diagnoses, enabling healthcare professionals to focus on patient care and treatment strategies.

AI and the intelligent workforce are revolutionizing healthcare in these key areas, and this is just the start. AI is enabling more personalized healthcare by analyzing patient data, including genetic information, to tailor treatments to individual patients. This approach is particularly transformative in areas like oncology, where treatments can be customized based on the genetic makeup of a patient's cancer.

AI systems are being used to predict patient risks by analyzing electronic health records (EHRs) and other data sources. This can identify patients at risk of chronic diseases or readmission, allowing for early intervention and better management of health conditions. AI accelerates the drug discovery process by predicting how different drugs will react in the body. This not only speeds up the development of new medications but also helps in repurposing existing drugs for new therapeutic uses.

AI-powered virtual assistants are increasingly used for initial patient engagement and triage, symptom checking, and provision of health information. Telemedicine, facilitated by AI, allows healthcare providers to diagnose and treat patients remotely, which is crucial for rural or underserved areas. AI tools assist healthcare professionals in making decisions by providing up-to-date medical information from journals, textbooks, and clinical practices. These systems can offer treatment recommendations based on the latest research and data.

Additionally, AI and virtual reality (VR) technologies are being used for the training and continuous education of medical professionals. They provide realistic simulations and can help in practicing surgical techniques or diagnosing virtual patients. It is being used in the detection and treatment of mental health disorders. For example, machine learning algorithms can analyze speech or typing patterns to detect signs of mental health issues, and chatbots can provide initial counseling and support. And AI is streamlining administrative tasks in healthcare, such as scheduling appointments, billing, and maintaining patient records, allowing healthcare professionals to focus more on patient care.

Overall, the collaboration between AI and human workers in healthcare is leading to a transformation in how diagnostics and treatments are delivered, emphasizing the complementary strengths of both, and aiming to improve patient outcomes while making the healthcare system more efficient.

EXAMPLE 2: CUSTOMER EXPERIENCE– ENGAGEMENT, SALES, AND SUPPORT

Another key example of how AI transforms industries is through improved customer experience engagement, sales, and support. By enhancing the efficiency, personalization, and responsiveness of customer interactions, AI can offer consumers a more user-friendly experience.

As one Salesforce article notes, "Customer service agents can realistically only handle one customer service call or problem at a time. AI, in its various incarnations, gives customer service departments the ability to do more, thus improving the customer experience."[45] AI-driven tools like chatbots and virtual assistants provide *immediate* customer assistance, handling routine inquiries and allowing human customer service representatives to focus on more complex issues. This improves the speed and convenience of support and ensures customers feel heard and supported.

In sales, AI is used to analyze customer data and behavior to make personalized product recommendations, which can lead to increased conversion rates and customer loyalty. AI-powered systems can predict which leads are more likely to convert, helping sales teams focus their efforts more effectively. Furthermore, AI enhances customer experience by providing insights into customer preferences and behaviors, enabling companies to tailor their strategies and create more meaningful interactions. By delivering personalized experiences, companies can build stronger customer relationships, leading to higher satisfaction and retention.

AI also plays a role in training and empowering the workforce. AI simulation roleplay and real-time data analysis can speed up the proficiency of new hires, especially in a hybrid workforce environment.

This ensures customer service agents are well-equipped to confidently handle a variety of customer interactions.

AI algorithms analyze customer data, including past purchases, browsing behavior, and preferences, to offer personalized recommendations and content. This personalization leads to a more engaging and satisfying customer experience, increasing customer loyalty and sales. AI-powered chatbots and virtual assistants provide twenty-four-seven customer service, efficiently handling inquiries, complaints, and FAQs. They can simultaneously manage a high volume of requests, ensuring quick response times and freeing human agents to handle more complex issues.

By leveraging predictive analytics, AI empowers businesses to anticipate customer behaviors and sales trends, optimize inventory management and targeted marketing, and identify new sales opportunities. Understanding customer needs and behaviors allows companies to proactively engage with their audience through relevant offers, enhancing customer satisfaction and loyalty.

> *AI is having and will have a profound impact on how customer care is delivered, as innovative brands will use the technology to improve personalization and the self-service journey for customers, which will increase lifetime value of those relationships.*
>
> ♀ **GARY ASH,**
> Chief revenue officer, Working Solutions

AI-driven systems enhance customer service efficiency by intelligently routing queries to the most suitable human agent, tailored to the complexity of the problem and the agent's specific skills. This alignment of customer needs with agent expertise results in quicker, more effective support. Furthermore, AI is expanding the capabilities of self-service options, like interactive voice response systems and knowledge bases, enabling customers to find solutions independently, thereby expediting issue resolution and enhancing user satisfaction.

The analysis of customer feedback and interactions through AI tools provides businesses with a clear picture of customer sentiments, helping them pinpoint areas for improvement in products, services, and customer interactions. In the e-commerce sphere, AI enhances the shopping experience by offering real-time assistance, product recommendations, and alerts on promotions based on a customer's browsing and purchase history.

AI's role extends to unifying various customer interaction channels such as email, social media, chat, and phone, ensuring a consistent and seamless experience across platforms. This integration is crucial in maintaining a high level of customer satisfaction. In addition, AI-driven marketing tools are revolutionizing how businesses automate and personalize marketing campaigns, leading to improved engagement and higher conversion rates.

EXAMPLE 3: FINANCE–ALGORITHMIC TRADING AND RISK MANAGEMENT

Third, AI and the intelligent workforce improve algorithmic trading and risk management by leveraging advanced machine learning algorithms to analyze vast amounts of market data at high speeds, identify patterns, and execute trades at optimal times. This enhances the ability to capitalize on market inefficiencies and manage investment portfolios more precisely.

In risk management, AI systems can simulate various scenarios to predict potential outcomes, aiding in informed decision-making. These systems are adept at monitoring real-time data to spot potential risks and prompt alerts for proactive risk mitigation. Internationally, banks leverage AI-powered natural language processing to scrutinize internal communications, spotting potential misconduct and ensuring trading compliance.

When it comes to trading, AI enhances both systematic and algorithmic strategies in several key areas. It can process extensive market data—including historical trends, economic indicators, and social sentiment—much more rapidly than humans, which supports timely and informed decision-making. AI's proficiency in discerning obscure market patterns aids traders in proactive investment decisions by predicting market shifts and identifying investment opportunities.

Moreover, AI's risk management is more nuanced, offering real-time assessments that enable traders to minimize potential losses and refine their strategies. AI-driven algorithms not only execute trades at the most favorable times, adapting to market changes, but also detect irregular trading patterns that could signal fraud, thus safeguarding financial integrity.

Compliance is another area where AI shines, ensuring that trading activities align with evolving regulations through automated monitoring, reporting, and auditing—a crucial aspect in the heavily regulated financial sphere. AI's ability to manage diversified portfolios tailored to individual risk profiles and rebalancing to maintain ideal asset allocation and exposure further showcases its significance in trading.

The advantage of AI extends to reducing human error in trading decisions, which can be costly, and executing trades in split seconds—key for high-frequency trading that capitalizes on slight market fluctuations.

Machine learning allows trading systems to evolve by learning from market data, enabling AI-driven trading strategies to adapt to market changes. The amalgamation of AI with human expertise signifies a more fortified approach to algorithmic trading and risk management, where AI handles data analysis and transaction processing and humans concentrate on strategic decision-making and cultivating customer relationships. This collaboration paves the way for sophisticated, data-driven decision-making that can mitigate risks while maximizing returns.

EXAMPLE 4: EDUCATION–PERSONALIZED LEARNING AND VIRTUAL TUTORS

This brings us to education. AI is poised to revolutionize the educational landscape, offering personalized learning experiences and reducing administrative burdens to allow educators to focus on teaching and mentorship.

Interactive simulations and AI tutors are at the forefront of this change, adapting to students' individual learning paces and styles to enhance understanding and retention of information. This personalized approach is particularly advantageous in online education and crowdsourcing services, improving the accuracy and efficiency of learning outcomes.

AI's capabilities extend beyond instructional assistance to automating administrative tasks such as grading and lesson planning. This shift enables teachers to allocate more time to student interaction and personalized instruction. Additionally, AI contributes to the development of digital microcredentials, facilitating lifelong learning and career development by documenting and translating skills across different contexts.

In anticipation of a changing job market influenced by AI and emerging technologies, the educational sector is adapting to prepare students with twenty-first-century skills, such as critical thinking, problem-solving, and teamwork. These skills remain in high demand despite the shift toward automation and the creation of new job roles that technology brings.

Further benefits of AI in education include tailoring content to match each student's needs, providing personalized support through tutoring systems, and creating engaging learning environments with gamification and simulations. AI tools also offer real-time language learning assistance, inclusivity for students with disabilities, and insights into job market trends to align curricula with industry demands.

Moreover, AI assists in the professional development of educators by analyzing teaching effectiveness and suggesting improvements. In higher education, AI aids in research by processing complex datasets and fostering discoveries. It also monitors students' mental health, providing crucial support when needed.

Overall, integrating AI into the educational process enhances the learning experience, leading to better educational outcomes and a well-prepared workforce for future challenges. This integration emphasizes the need for a balanced approach where AI complements human expertise, allowing for a comprehensive and adaptive educational environment.

EXAMPLE 5: TRANSPORTATION–AUTONOMOUS VEHICLES AND TRAFFIC MANAGEMENT

The final example we'll include is transportation.

AI is set to revolutionize transportation, especially with its application in autonomous vehicles (AVs), offering significant enhancements in safety, efficiency, and convenience. AI algorithms process data from an array of sensors and cameras to navigate roads, recognize obstacles, and make critical decisions in fractions of a second. This reduces accidents caused by human error and promises a safer driving experience.

In the broader transportation sector, AI's impact is multifaceted. It is integral to traffic management systems, optimizing traffic flow and reducing congestion through real-time adjustments to traffic signals and route recommendations. Predictive maintenance, powered by AI's sensor data analysis, ensures vehicle reliability and safety by anticipating maintenance needs before breakdowns occur.

AI extends its benefits to electric autonomous vehicles by optimizing battery usage and driving patterns, which enhances energy efficiency and lowers the carbon footprint of transportation. In public transit and ride-sharing services, AI personalizes the passenger experience, offering route recommendations and real-time travel updates while also streamlining ride-sharing options to improve efficiency.

Regulatory compliance is another area where AI assists manufacturers by ensuring that autonomous vehicles operate within safety and legal guidelines. The data gathered by AI-enabled vehicles is invaluable for urban development, informing smarter infrastructure planning.

Communication between vehicles (V2V) and road infrastructure (V2I) is enhanced through AI, promoting safety and efficiency on the roads. In logistics, AI-driven route optimization for freight transport reduces delivery times and costs, while advanced security features in autonomous vehicles, such as real-time monitoring and incident reporting, bolster overall safety.

> At Siemens, our purpose is to create technology to transform the everyday for everyone and AI is a critical part of this future. Whether it is utilizing AI to make our internal processes more efficient or harnessing its power in our software to optimize rail

operations, AI supplements human ingenuity by making what we do faster and more efficient.

♀ **MARC BUNCHER,**
President and CEO, Siemens Mobility

AI's influence extends beyond ground vehicles, with autonomous drones emerging as a fast and efficient option for transporting goods, especially in last-mile logistics. The technology's capacity to improve traffic management systems and optimize logistics and fleet management is also noteworthy, leading to more effective delivery and transportation services.

Overall, AI is pivotal for advancing autonomous technology and fostering smarter, more integrated transportation networks. It underpins the ongoing modernization of transport infrastructure, enhancing safety and sustainability and playing a crucial role in the development of smart city initiatives. As AI continues to develop, its collaboration with human oversight promises a future of strategic, data-driven transportation solutions.

A Look at Where We Are Headed

One of the latest AI innovations in the news combining humans and machines is the brain-computer interface, called Neuralink. Neuralink is a neurotechnology company founded by Elon Musk and others in 2016, with the goal of developing brain-computer interfaces (BCIs) to connect the human brain directly to computers. Neuralink's mission is to enable humans to communicate with computers and other electronic devices using only their thoughts, potentially revolutionizing how we interact with technology.

The company is working on highly advanced BCI technology that involves implanting tiny threads—thinner than a human hair—into the brain. These threads are connected to an external device, known as the Link, which serves as the communication bridge between the brain and external devices or computers. The technology aims to read signals from the brain with high precision, bandwidth, and resolution, and to transmit these signals to computers for various purposes.

One of the primary goals of Neuralink is to address and treat neurological conditions and disabilities. The technology could potentially help people with paralysis to control electronic devices, restore sensory and motor functions, and even treat neurological disorders like Parkinson's disease, epilepsy, and depression.

Beyond medical applications, Neuralink's technology could one day enable humans to enhance their cognitive abilities, such as faster information processing, memory enhancement, and direct interfacing with digital information and artificial intelligence. The technology could revolutionize how we communicate, allowing for thought-to-text or thought-to-speech communication, bypassing traditional physical interfaces like keyboards and touchscreens.

The advent of AI and intelligent automation is reshaping the industrial landscape, leading to the emergence of "superjobs." These roles synergize the strengths of both humans and machines, integrating AI's computational power with human skills like emotional intelligence, ethical judgment, and creative conversation design. This fusion is altering job functions and enhancing the value of human workers.

But while many are inclined to see this as a bad thing, the collaboration between humans and machines can be powerful. In their Deloitte article, Gina Schaefer and Ryan Sanders write, "While the idea of machines 'taking over the world' makes for compelling science fiction, businesses are discovering that the most powerful use cases for intelligent automation (IA) involve people and machines working together as a team."[46]

The concept of "human in the loop" is gaining traction, highlighting the indispensable role of human expertise in training, supervising, and guiding AI systems, and in driving creative decision-making processes. Despite increasing automation, this approach underscores the indispensable role humans play in the modern workflow.

Thankfully, many companies are proactively preparing their workforce for this shift, focusing on reskilling and upskilling to ensure seamless collaboration between employees and AI tools. This educational investment is crucial for harnessing the benefits of AI and intelligent automation in the workplace. However, the challenge lies in managing a hybrid workforce of humans and AI. Leaders must candidly address the concerns that arise with such integration to prevent unease and distrust. Clear and factual communication is key to replacing rumors and fears with understanding and acceptance.

Intelligent automation (IA) should be positioned as an *aid* to human workers, not a *replacement*. It is intended to relieve employees of monotonous tasks, allowing them to concentrate on more strategic and fulfilling work. When implemented thoughtfully, IA can lead to enhanced job performance and greater job satisfaction.

To summarize, as machine learning, natural language processing, computer vision, and robotics continue to evolve, they are driving a profound transformation across various industries. This evolution necessitates a strategic approach to workforce development and reimagining work in the modern enterprise, focusing on augmenting human capabilities, boosting productivity, and inspiring innovation.

CHAPTER 7:

CHALLENGES AND ETHICAL CONSIDERATIONS

THE ETHICS OF **AI**

AS AI BECOMES MORE INGRAINED IN OUR DAILY ROUTINES, tackling ethical issues becomes a top priority. Unfortunately, many companies are ill-equipped to tackle this challenge. Jason Furman writes, "The problem is these big tech companies are neither self-regulating, nor subject to adequate government regulation."[47] This is a major problem to consider.

Like a double-edged sword, AI has the potential to bring immense benefits and opportunities, such as increased efficiency, productivity, and convenience. However, it also carries inherent risks and ethical challenges. Just as a double-edged sword must be wielded carefully to avoid harm, AI must be developed, deployed, and regulated with careful consideration for its potential negative consequences, such as bias, privacy infringement, and job displacement.

Introducing AI and automation tech, especially in reshaping our digital workforces, raises a slew of ethical dilemmas. From concrete worries like safeguarding data privacy to fuzzier ones like algorithmic bias, it's a minefield. Navigating this terrain calls for a nuanced approach, considering tech specs, rules, and societal values.

Dealing with these ethical hurdles goes beyond mere rule-following; it demands a holistic view. It's critical to weave fairness, transparency, and accountability into the very fabric of AI development and deployment. This sets the stage for a regulatory framework built on trust and responsibility.

When discussing AI, ethical considerations come to the forefront, accompanied by intense moral deliberations. Thus, it is imperative to establish and adhere to ethical guidelines governing AI, encompassing principles such as fairness, accountability, transparency, and privacy. Moreover, it is essential to address the broader ethical implications, ensuring inclusivity and carefully considering the societal consequences as AI systems are implemented.

"AI is not neutral: AI-based decisions are susceptible to inaccuracies, discriminatory outcomes, embedded or inserted bias."[48]

The Challenge of Bias and Discrimination

When AI systems pick up biases, it's like inheriting a family trait. Only in this case, it's societal prejudices. These biases can creep into everything from job applications to predictive policing, making AI not just inaccurate but also unfair and even discriminatory.

Take, for instance, an AI trained on data from a company where men mostly landed the jobs. If left unchecked, it might unfairly favor male candidates in future selections. The same holds true for racial biases. Karen Mills, senior fellow at Harvard Business School and head of the US Small Business Administration from 2009 to 2013, states, "If we're not thoughtful and careful, we're going to end up with redlining again."[49]

Sometimes, AI can create its own little feedback loops, like a self-fulfilling prophecy. Imagine a predictive policing tool placed in a neighborhood with a history of high arrest rates. It could end up sending more cops there, leading to more arrests, not necessarily because there's more crime, but simply because there's more surveillance. And guess what? That feeds right back into the AI, reinforcing its bias.

"Part of the appeal of algorithmic decision-making is that it seems to offer an objective way of overcoming human subjectivity, bias, and prejudice," says political philosopher Michael Sandel. "But we are discovering that many of the algorithms that decide who should get parole, for example, or who should be presented with employment opportunities or housing ... replicate and embed the biases that already exist in our society."[50] Sandel goes on to say, "AI not only replicates human biases, it confers on these biases a kind of scientific credibility. It makes it seem that these predictions and judgments have an objective status."[51]

Mitigating bias is like putting on corrective lenses for AI developers. Without diverse teams and perspectives, they might miss those subtle biases lurking in the data or the algorithms. And it's not just the developers; how we interpret AI outputs matters too. Even if the AI itself is pretty unbiased, if we view its results through our own skewed lenses, we're just adding fuel to the fire of social biases.

So what's the solution? It starts with diversifying the data that AI learns from, making sure it's not just reflecting one slice of society. Then, there's the design of the AI models themselves. Every little choice—from what factors to include to how heavily to weigh them—can either dial up or dial down the biases.

Data Privacy and Security

In the digital age, where cloud-based platforms reign supreme, our worries about data privacy and security have hit an all-time high. As Forbes Business Council member Jackie Shoback writes, "In today's business environment, data is one of the most valuable assets a company possesses."[52] Now toss in the era of artificial intelligence, and suddenly it's not just about who has access to our cat videos but also about the very essence of our personal information being at stake. We're at a crossroads where groundbreaking innovation meets the potential dark alley of privacy invasion.

Artificial intelligence models, sophisticated constructs essential for various applications ranging from voice assistants to predictive analytics, demonstrate an unquenchable thirst for data. Consequently, they emerge as sought-after entities in the digital landscape, attracting the attention of hackers. Given their reliance on extensive datasets for training and functioning, these AI models not only advance technological boundaries but also expose significant risks to privacy and security, akin to opening Pandora's box.

The foremost of these risks is cyberattacks. It's not just about some lines of code being exposed; it's about your sensitive information doing an unscheduled world tour without your permission. Your data could end up being misused or manipulated, or worse, landing in the wrong hands. The sheer collection of these massive datasets, often pulled from the digital footprints of individuals like you and me, raises eyebrows about just how much of our lives are on display without us even knowing it.

Now, here's the kicker—the pace at which AI technology is sprinting forward leaves regulatory frameworks panting to catch up. Our laws, well-intentioned as they may be, find themselves in a perpetual game of catch-up. Enter General Data Protection Regulation (GDPR), the superhero of data protection laws. But even compliance with this mighty shield becomes tricky when the AI landscape is constantly shifting.

So what's a savvy user to do? It's all about taking back the reins. Incorporating robust data security measures is like putting a moat around your digital castle. Adhering to data protection regulations, like following the sacred commandments of GDPR, gives you a say in how your information is handled. And let's not forget the secret weapons—data anonymization and encryption. These techniques add an extra layer of armor, making sure your data stays yours and yours alone.

In this digital Wild West, a bit of vigilance and the right tools can make all the difference in keeping your digital homestead safe and sound.

Transparency, Responsibility, and Accountability

The lack of transparency in AI systems is a pressing issue that reverberates across various sectors, casting a shadow over the adoption and effective utilization of AI technologies. Particularly concerning are AI models, especially those underpinned by deep learning algorithms, often likened to "black boxes" due to their inscrutable internal mechanisms. This opacity obstructs comprehension of how these models arrive at decisions, rendering them enigmatic and challenging to scrutinize.

One of the most troubling implications of this opacity is its potential to conceal biases ingrained within AI systems. Without visibility into the decision-making processes, biases can lurk undetected, perpetuating unfair or discriminatory outcomes. This opacity not only complicates efforts to uncover and rectify biases but also erodes trust among users and stakeholders, especially in critical domains like healthcare, finance, and law enforcement, where the consequences of erroneous decisions can be profound.

But as one Deloitte article states, *transparent* AI is *explainable* AI.[53] This means the language of AI has to be broken down so everyday users understand the fundamentals of how it operates. Stefan van Duin writes, "AI is smart, but only in one way. When an AI model makes a mistake, you need human judgment. We need humans to gauge the context in which an algorithm operates and understand the implications of the outcomes."[54] In other words, we need accountability.

Unfortunately, the lack of transparency in AI systems engenders ambiguity regarding accountability. In instances where AI-driven decisions yield negative outcomes, determining responsibility becomes a convoluted endeavor. This absence of clear accountability can have far-reaching legal and ethical ramifications, raising questions about liability and exacerbating distrust in AI technologies.

Opaque AI systems pose a tough challenge when it comes to being fair and accountable. Without transparency, figuring out why biased outcomes happen and fixing them becomes really hard. That's why it's superimportant for AI systems to be clear and explainable, so people can understand why decisions are made. To make AI transparent, we need to use different methods like model interpretability. This helps to uncover how AI systems make decisions. Also, providing clear info about what AI can and can't do is key for people to trust and understand it better.

Transparency is a moral obligation. By shedding light on how AI works and being honest about its strengths and weaknesses, we can tackle the ethical challenges that come with using AI responsibly.

The Moral Dilemma

Let me share an illustration to highlight the moral challenges we face.

Once upon a time, in a bustling city, there was a self-driving car company named AutoDrive. AutoDrive was known for its advanced AI systems that powered its autonomous vehicles. One day, they faced a moral dilemma that challenged their AI's decision-making capabilities.

During a routine test drive, one of their self-driving cars encountered an unexpected situation. A group of pedestrians suddenly crossed the road, ignoring the red light. At the same time, a motorcyclist was approaching from the opposite direction, exceeding the speed limit. The AI system had to make a split-second decision: should it swerve to avoid the pedestrians, potentially hitting the motorcyclist, or should it continue straight, risking the lives of the pedestrians?

This situation posed a significant ethical challenge for AutoDrive. The AI system was programmed to minimize harm, but in this scenario, any decision it made would potentially harm someone. The company had to grapple with the question of whose safety should be prioritized in such situations. This moral dilemma sparked a debate within AutoDrive and the wider community. Some argued that the AI should prioritize the safety of the pedestrians, as they were more in number. Others contended that the motorcyclist, despite speeding, had the right of way and should be protected.

In response to this incident, AutoDrive decided to review its AI systems and ethical guidelines. They involved ethicists, social scientists, and philosophers in the conversation to ensure a more comprehensive approach to ethical AI development. They also started a dialogue with their users and the public to understand their views on such moral dilemmas.

This story serves as a reminder of the ethical considerations that need to be addressed in the design and deployment of AI systems. It underscores the importance of transparency, human oversight, and continuous feedback in ensuring that AI systems align with human values and societal norms.

Societal Impact

The advent of AI brings with it a wave of transformative changes with far-reaching implications across various spheres of society. One of the foremost concerns is the potential for job displacement across different sectors. As AI systems automate tasks traditionally performed by humans, there is a looming fear of widespread unemployment. This displacement can disproportionately affect certain industries and demographics, exacerbating existing socioeconomic inequalities.

Moreover, the digital divide stands out as a critical issue in the age of AI. While advancements in AI have the potential to drive innovation and economic growth, they may widen disparities between those who have access to and benefit from AI technologies and those who do not. This digital gap could deepen existing social inequalities, further marginalizing vulnerable populations and exacerbating divisions within society.

The ethical implications of AI extend beyond the scope of employment and access to technology. In the context of warfare, the development and deployment of autonomous weapons raise profound ethical dilemmas. The prospect of AI-enabled warfare blurs the lines between human agency and technological autonomy, raising concerns about accountability, proportionality, and the protection of civilians in armed conflict.

Addressing these challenges requires a multifaceted approach. Job displacement resulting from AI adoption necessitates concerted efforts toward reskilling and upskilling the workforce. Initiatives aimed at providing training and education on the ethical use of AI are crucial to ensuring that employees are equipped to navigate the ethical complexities of AI technologies. Moreover, ongoing public awareness campaigns are essential to fostering informed discourse and understanding about the benefits and risks of AI across society.

The Regulatory and Legal Landscape

As the influence of AI continues to permeate various aspects of society, the regulatory landscape surrounding its governance is under scrutiny. Existing regulations aimed at addressing AI challenges vary in their adequacy and effectiveness. While some jurisdictions have established comprehensive frameworks to govern AI development and deployment, others lag behind, leaving regulatory gaps and uncertainties. As Associate General Counsel for HireRight Alonzo Martinez writes,

> Employers understand that AI can significantly enhance the efficiency and precision of various aspects of the employment process. It offers the potential for substantial gains in terms of

speed and accuracy. However, it's equally important to strike a balance between leveraging AI for its capabilities while also ensuring that the technology aligns with legal requirements and maintains fairness and non-discrimination in employment decisions.[55]

When it comes to regulating AI, different countries have their own perspectives. Some are keen to push the boundaries of tech and foster innovation, while others are more focused on safeguarding privacy, human rights, and the overall well-being of society. These differences reflect a mix of cultural, political, and economic factors, shaping how we talk about ethics and rules for AI worldwide.

But one thing we know for sure: the legal challenges and complexities of AI will only grow in the coming years. For companies diving into AI, following the rules is key. Keeping up with the latest laws and regulations on AI is crucial to stay on the right side of the law and dodge legal trouble. That means understanding how AI decisions can have legal consequences, like the need to explain decisions under rules like GDPR.

The Path Forward

Addressing the challenges and ethical dilemmas posed by AI requires collaboration, especially through public-private partnerships. By bringing together experts from various disciplines like technology, ethics, sociology, and psychology, we can approach AI ethics comprehensively. Educating both users and developers about ethical guidelines is crucial for fostering a culture of responsibility and integrity in AI development and deployment.

Monitoring and evaluating AI systems continuously is essential to ensure ethical adherence. By regularly assessing their performance and impact, we can identify and address ethical concerns early on, minimizing potential harm. Also, we must carefully weigh the ethical implications of AI advancements to ensure they align with societal values, covering aspects like privacy, security, fairness, and transparency.

Ethical responsibility in AI demands holding technology to a higher standard, promoting human rights and societal well-being, and avoiding biases or inequalities.

This entails not only designing AI systems with ethics in mind but also being accountable for their impact on individuals and communities. Transparency is crucial for building trust in AI; without clear explanations of AI operations and decision-making, trust diminishes, and ethical concerns arise. Ultimately, prioritizing transparency and explainability in AI development and deployment fosters trust and confidence in these technologies, ensuring they serve society responsibly and ethically.

CHAPTER 8:

THE FUTURE OF THE INTELLIGENT WORKFORCE

SO WHAT DOES THE FUTURE OF THE INTELLIGENT WORKFORCE look like? As one McKinsey article notes, "The future of work is already here, and it's moving fast."[56]

In this new era, it's important to reemphasize that AI and automation don't have to be job stealers; they're tools that *enhance* what humans can do. People team up with intelligent systems to get more done, be more creative, and make better decisions. This partnership lets humans handle the tricky, strategic, and social parts of their jobs, while AI takes care of the routine, number-crunching, or risky tasks.

The relationship between humans and machines is a bit like an orchestra. In an orchestra, each musician plays a unique instrument, contributing their own skills and expertise to create a harmonious and cohesive performance. Similarly, in the intelligent workforce of the future, humans and machines will work together in harmony, leveraging their respective strengths to achieve common goals.

Just as each section of the orchestra has its role to play in producing beautiful music, different departments and functions within organizations will collaborate seamlessly, with humans and AI technologies complementing each other's abilities. Like the conductor guiding the orchestra, leaders and managers will oversee and coordinate the efforts of both humans and machines, ensuring that they work together toward a shared vision.

Following the orchestra theme, just as a well-rehearsed orchestra can adapt to changes in tempo, dynamics, and style, the intelligent workforce will demonstrate agility and flexibility in response to evolving business needs and technological advancements. Humans will provide creativity, intuition, and emotional intelligence, while AI technologies will offer speed, accuracy, and data-driven insights, resulting in a dynamic and responsive organizational culture.

As industries keep evolving with AI, machine learning, automation, and digital teamwork tools, keeping our skills up to date is a must. Lifelong learning and staying on top of new tech are key to staying competitive in the ever-changing job scene. With online courses, company training, and other educational programs, your team can keep learning and adapting to new job demands, staying relevant in the shifting work landscape.

This process is known as reskilling.

How Companies Are Reskilling Their Workforces

Reskilling for AI refers to the process of acquiring new skills or updating existing ones to adapt to the integration of AI technologies in the workplace. Here are several practical ways your company can navigate the reskilling journey.

First off, it's about spotting the gaps. You need to identify the skills that will be in demand as AI becomes more prevalent. This means understanding how AI will impact your current roles and processes and pinpointing the new skills that will be needed. Think data literacy, AI ethics, machine learning—the works. Since many of these skills aren't covered in traditional education or older training programs, there's a bit of a gap that needs filling through reskilling efforts.

As Ian Bird writes in his IBM article, "The future of reskilling, training and talent development uses AI to provide employees with less prescriptive courses by taking their experience, role and existing skills into account to offer a truly personalized and relevant learning pathway unique to them along with greater learning in the flow of their work."[57]

Next, it's all about personalized learning. You can leverage AI to tailor learning experiences to you. By analyzing your current skills, learning pace, and preferences, AI can recommend customized training programs. Platforms like Coursera, Udemy, and LinkedIn Learning offer a plethora of courses on AI, data science, and related fields, making it easy for your company to provide you with access to relevant training.

Creating a culture of continuous learning is also key. Your company can encourage ongoing learning by setting aside time during work hours, offering incentives for completing courses, and fostering an environment where knowledge sharing is the norm.

Collaboration is another biggie. Some companies team up with universities and technical schools to develop courses tailored to their specific needs. These partnerships can even lead to employees earning accredited qualifications. And let's not forget about mentorship programs and peer learning groups—they're invaluable for passing on knowledge and skills within the organization.

Last, it's about staying ahead of the curve. You need to keep an eye on emerging trends and technologies to anticipate future skill requirements. This proactive approach ensures that your workforce is ready for whatever the future throws its way. Reskilling for the age of AI isn't just about keeping up with the latest tech. It's about building a workforce that's adaptable, innovative, and ethical in its use of AI. By investing in learning and development and fostering a culture of continuous learning, your company can not only stay competitive but also support you through the shifts brought about by AI and automation.

Integrating AI in Workspaces

Integrating AI in workspaces is like weaving a tapestry of technology into the fabric of your organization. Just as each thread contributes to the overall design, AI seamlessly intertwines with your existing workflows, enhancing their strength and beauty. Like a skilled weaver, you carefully select and blend different AI tools and algorithms, creating a harmonious composition that supports and enriches your business processes.

As you weave AI into the fabric of your workspace, you may encounter moments of uncertainty or resistance, much like a weaver adjusting to unforeseen knots or tangles in the thread. However, with patience and skill, you untangle these complexities, transforming them into opportunities for learning and improvement.

Integrating AI into your workplace represents a fundamental shift in how you operate. It's not just about implementing new technology; it's about transforming your entire organizational culture, redefining processes, and upgrading skill sets. In this new era, the synergy between humans and machines is paramount. While AI, machine learning, and automation offer unprecedented capabilities, they're most effective when they work in tandem with your human ingenuity and expertise.

The future of work will be collaborative, with you and your machines complementing each other's strengths. As CEO of Value-matrix.ai, Aditya Malik, writes, "Integrating AI into society should be a collaborative and co-evolving process rather than an imposition of new technologies."[58]

Advanced AI algorithms can analyze vast amounts of data, identify patterns, and generate insights that you may overlook. By leveraging AI-powered tools, you can make more informed decisions, streamline processes, and drive business growth. Moreover, AI can automate routine tasks, allowing you to allocate your time and energy to tasks that require your human judgment and creativity.

Machine learning algorithms further enhance the collaborative dynamic between you and your machines. These algorithms can learn from data, adapt to changing circumstances, and improve their performance over time. As you interact with machine learning systems, you provide feedback and guidance, enabling these systems to continuously refine their models and deliver more accurate results. This iterative process of learning and improvement empowers you to stay agile and responsive in an ever-evolving business landscape.

Automation, meanwhile, streamlines repetitive tasks and workflows, reducing human error and accelerating processes. By automating routine tasks, you can focus on value-added activities that require your human intervention, such as customer interaction, innovation, and strategic decision-making. This not only boosts your productivity but also enhances your job satisfaction and engagement.

However, the successful integration of AI into your workplace requires more than just technological implementation. It necessitates a cultural shift toward embracing innovation, experimentation, and lifelong learning. You must foster a culture of continuous improvement, where you feel empowered to adapt to new technologies, acquire new skills, and explore innovative solutions to business challenges. Additionally, you must invest in training and development programs to upskill your workforce and ensure it has the expertise to leverage AI effectively.

SIX PRACTICES FOR INTEGRATING AI INTO THE WORKFORCE

There are six primary practices for integrating AI into your workforce, each essential for navigating the complexities of AI adoption effectively. By following these practices, you can harness the power of AI to drive innovation, improve productivity, and foster a culture of continuous improvement within your workforce.

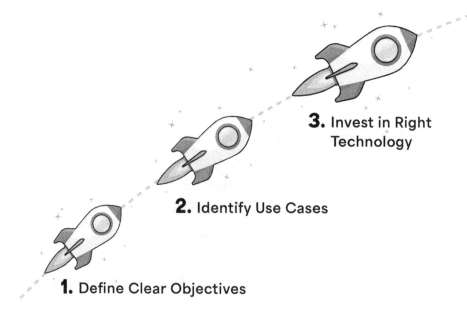

3. Invest in Right Technology

2. Identify Use Cases

1. Define Clear Objectives

PRACTICE 1: DEFINE CLEAR OBJECTIVES

Establishing clear objectives is the cornerstone of successful AI integration. By defining specific goals such as improving productivity, enhancing customer service, or streamlining operations, you can align your AI initiatives with strategic priorities. Starting small with pilot projects allows for experimentation and learning without committing significant resources up front. These pilot projects serve as test beds to demonstrate the value of AI in real-world scenarios and allow for adjustments and refinements before scaling up. By setting clear objectives and starting with manageable projects, you can mitigate risks and ensure a more successful AI implementation.

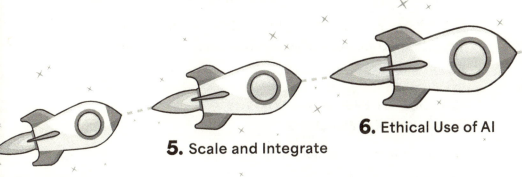

6. Ethical Use of AI

5. Scale and Integrate

4. Workforce Skill Enhancement

PRACTICE 2: IDENTIFY USE CASES

Identifying the right use cases for AI is crucial for maximizing its impact. You must pinpoint areas where AI can add the most value, whether it's automating routine tasks, improving data analysis, or enhancing customer service with chatbots. Leveraging pilot programs before a full-scale rollout is essential, as AI integration often begins with small, controlled implementations. These pilot projects provide valuable insights into the technology's impact and allow you to refine your approach based on real-world feedback. By carefully selecting use cases and starting with pilot projects, you can ensure that your AI initiatives deliver tangible benefits and align with strategic objectives.

PRACTICE 3: INVEST IN THE RIGHT TECHNOLOGY

Investing in the appropriate AI technologies is also critical for success. You must choose the right tools and platforms based on your specific use cases and requirements. This could involve adopting off-the-shelf AI software, developing custom solutions in-house, or partnering with specialized vendors. Additionally, you must ensure you have robust data infrastructure in place to support AI systems. This includes collecting, storing, and processing data while adhering to data privacy and protection standards. By investing in the right technology and data infrastructure, you can lay the foundation for successful AI integration and drive business value.

PRACTICE 4: DEVELOP WORKFORCE TRAINING AND RESKILLING

Ensuring that your employees are equipped to work alongside AI tools is essential for successful integration. This requires providing technical training for IT staff and educating other employees on how to interact with and leverage AI in their roles. Moreover, fostering a culture of innovation, adaptability, and continuous learning is crucial for successful AI integration. Employee buy-in is essential, as there can be resistance to AI due to fears of job displacement or mistrust in technology. By investing in workforce training and fostering a culture of continuous learning, you can ensure that your employees are prepared to embrace AI and drive its successful implementation.

PRACTICE 5: SCALE AND INTEGRATE

Scalability and integration are key considerations for long-term AI success. Once AI applications prove successful in pilot projects, you can scale and integrate these solutions more broadly across your organization. This requires ensuring that AI systems can interact seamlessly with existing IT infrastructure. AI integration is an ongoing process that requires regular updates and adaptations as technology evolves. By focusing on scalability and integration, you can ensure that your AI initiatives remain effective and aligned with business objectives.

PRACTICE 6: ENSURE ETHICAL AND RESPONSIBLE USE OF AI

Establishing guidelines for ethical AI use is essential for building trust and ensuring responsible AI integration. This includes focusing on fairness, transparency, accountability, and privacy in AI algorithms and decision-making processes. Additionally, you must address potential biases in AI algorithms and ensure that AI-driven decisions are explainable and responsible. By prioritizing ethical considerations, you can build trust with employees, customers, and other stakeholders and ensure that your AI initiatives deliver positive outcomes for all parties involved.

Integrating AI into your workplace requires a comprehensive approach that goes beyond just technical implementation. It involves careful planning, a focus on data, infrastructure, workforce engagement, training, and attention to ethical considerations. By following these best practices, you can integrate AI in a way that enhances the work environment, drives business value, and ultimately leads to a more productive and engaged workforce.

The Next Big Disruptions

Predicting the next major disruption *after* AI is challenging, as it involves speculation about future technological advancements and their societal impacts. However, several emerging technologies and trends have the potential to be highly disruptive in the coming years, with the main focus being quantum computing, which will exponentially increase the pace of innovation.

As one *Forbes* article notes, "Quantum computing is poised to make exponential changes to a multitude of functions across industries and entities, which could radically change how we live and work."[59] Quantum computers, which operate on the principles of quantum mechanics, have the potential to solve complex calculations and problems at unprecedented speeds compared to current computers. This could revolutionize fields like cryptography, drug discovery, financial modeling, material science, and climate modeling. Quantum computing alone will create new job roles, requiring a workforce skilled in quantum algorithms and programming.

Advances in biotechnology, particularly in genomics and gene editing (like CRISPR technology), will lead to significant breakthroughs in medicine, agriculture, and bioengineering. This includes personalized medicine, genetically modified organisms for food security, and potentially curing genetic diseases.

Nanotechnology and the manipulation of matter at an atomic or molecular scale will transform various industries by enabling the production of new materials with specific, customized properties. This could impact manufacturing, healthcare (through targeted drug delivery systems), and energy storage. Neurotechnology and Brain-Computer Interface (BCI) technologies that can interface directly with the human brain present profound possibilities for enhancing cognitive abilities, treating neurological disorders, and even merging human consciousness with machines.

Robotic process automation (RPA) is the physical embodiment of AI in robots and autonomous systems that can revolutionize industries, from manufacturing to logistics, and even personal assistance. By taking over these tasks, RPA can free up employees to focus on more strategic, creative, and complex problem-solving activities, thus elevating the nature of work and potentially increasing job satisfaction. RPA can also lead to the creation of new roles in digital bot development, management, and maintenance, as well as in process reengineering to optimize the use of automation.

One topic that needs to be addressed when discussing the future of AI is singularity, a hypothetical future point where AI surpasses human intelligence, leading to unprecedented technological growth and societal change. The concept of singularity will have profound implications for the intelligent workforce. While the exact nature and timing of the singularity are subjects of speculation, its potential impact on the workforce can be understood by witnessing AI's results for accelerating innovation and productivity.

The core idea of the technological singularity is the development of artificial intelligence that is not just slightly more intelligent than humans but vastly more so, creating a superintelligence. This superintelligent AI would be capable of self-improvement and recursive enhancement, leading to an intelligence explosion where its capabilities could rapidly exceed all human intelligence.

The timeline for reaching the singularity is highly speculative, with predictions varying widely among experts. Some suggest it could happen within the twenty-first century, while others believe it may take longer or that it may not happen at all. The uncertainty stems from the complex challenges in creating AI that can fully replicate or surpass human cognitive abilities.

Given the potential for both positive outcomes and significant risks, many researchers advocate for careful and ethical development of AI technologies. This includes establishing safety protocols, fostering international cooperation, and engaging in multidisciplinary research to ensure that advances toward superintelligence are aligned with human interests.

All these areas carry their own set of opportunities, challenges, and ethical considerations. The extent and nature of their disruption will depend on technological breakthroughs, market dynamics, regulatory frameworks, and social acceptance. As with AI, the key will be to balance innovation with careful consideration of the broader impacts of AI on society and the environment.

The Future of Innovation

It's important to remember that all future innovation not only represents technological advancements but also necessitates a rethinking of current job roles, the creation of new types of employment, and a focus on continuous learning and adaptation. The workforce of the future will likely need to be more agile, with an emphasis on skills like adaptability, problem-solving, and lifelong learning.

Quantum computing, with its potential to perform complex calculations at unprecedented speeds, is poised to have a significant impact on the workforce. Its effects will be felt in various ways, like the creation of new job roles, such as algorithm developers, quantum hardware engineers, and quantum information theorists. It will also drive the demand for specialized skills in quantum mechanics, computer science, and related fields. Educational institutions and training programs will need to adapt to provide the necessary training and qualifications.

Like other advanced technologies, quantum computing will have a significant impact on the workforce. Preparing the workforce for quantum computing is a challenge, as it requires highly specialized knowledge that is currently not widely taught outside of academia. It is expected to automate more tasks currently done by humans, leading to changes in occupations and the skills required.

Some occupations may decline, while others will grow, and many will evolve. Workers will need to acquire new skills and adapt to working alongside increasingly capable machines. Demand for advanced technological skills such as programming will grow rapidly. Social, emotional, and higher cognitive skills, such as creativity, critical thinking, and complex information processing, will also see growing demand.

The overall labor market will have higher demand for social-emotional and digital skills. Although the demand for basic cognitive and manual skills is likely to decline, physical work is not going away. It may still account for just under 31 percent of time spent, driven by growth in sectors such as transportation services, construction, and healthcare.

The workforce will need to be more qualified and skilled than ever before. Just as the Industrial Revolution automated the manufacturing industry, creating specialized machine operators, the Knowledge Revolution will require specialized skill sets to complete the tasks and run the AI technology machines of the future.

Workers in lower-wage jobs are up to fourteen times more likely to need to change occupations than those in higher-wage positions, with women being 1.5 times more likely to need to move into new occupations than men[60] and most needing to learn additional skills to do so successfully. And by 2030, there will be an estimated 23 percent increase in the demand for STEM jobs.[61]

Although layoffs in the tech sector have been making headlines, this does not change the longer-term demand for tech talent among companies of all sizes and sectors as the economy continues to digitize. Policymakers, educators, and industry leaders will need to collaborate to ensure that the workforce is prepared for these changes.

The future intelligent workforce is also expected to be more technologically integrated, flexible, and dynamic, with a strong emphasis on human-AI collaboration, continuous learning, and adaptation to new roles and skills. This future will not only transform how work is done but also redefine the skills and attributes valued in the workplace. This shift will require new approaches to management, collaboration, and workforce engagement to ensure that employees remain motivated and productive in a digital-first environment.

The intelligent workforce of the future will be characterized by its ability to adapt, learn, and innovate, leveraging the strengths of both humans and machines to address complex challenges and drive growth in various sectors.

It's up to you to be ready for this change.

CHAPTER 9:

SUCCESS STORIES OF HUMANS AND MACHINES

IN THIS CHAPTER, LET'S DIVE INTO SOME REAL-LIFE EXAMPLES that show how AI supercharges businesses across different areas and combines AI's data processing speed with human emotional intelligence, creativity, and decision-making skills.

AI is revolutionizing different industries, including healthcare, finance, and customer service. In healthcare, AI helps doctors diagnose patients faster and more accurately by analyzing a vast amount of medical data. However, doctors still have a crucial role in bringing their experience and intuition to the table. Together, AI and doctors work as a powerhouse duo to ensure that patients receive top-notch care tailored to their needs.

In finance, AI is like a financial wizard, crunching numbers and spotting trends in the stock market. But it's not infallible. That's where human investors come in, using their gut instincts and knowledge of the market to make smart decisions. With AI and humans working side by side, investment firms can stay ahead of the game and deliver big wins for their clients.

AI chatbots are becoming more common in customer service, making it easier for companies to handle customer inquiries. However, human agents bring warmth and understanding to the table that AI can't match. Whether it's a tricky problem or just a friendly chat, human agents provide the personal touch that customers crave. Together, AI and humans create a customer service dream team that keeps clients happy and coming back for more.

> *I envision the next decade of human-machine collaboration as a catalyst for unprecedented efficiency and personalization across industries, with a notable impact in fields that we service, such as real estate. AI and machine learning will revo Humach lutionize how we access, manage, and interact with services, making personalized experiences the norm rather than the exception. By leveraging conversational AI, we can offer real-time, tailored advice to buyers and renters, automate routine inquiries, and provide virtual tours, significantly enhancing the client experience.*
>
> **♀ MASON LEVY,**
> Chief executive officer, Swivl

Together, AI and humans can achieve amazing things.

Humach and a Leading Theme Park and Entertainment Company

Let's look at a specific example.

The use of digital agents collaborating with human specialists is often referred to as "hybrid AI systems" or collaborative intelligence. These systems combine the strengths of human intelligence, such as emotional understanding, creativity, and complex decision-making, with the speed, scalability, and data-processing capabilities of artificial intelligence.

Humach, the company I lead, is a customer experience provider that focuses on combining human talent and advanced AI technology to create tailor-fit customer service experiences. The term *Humach* is a portmanteau of "human" and "machine," reflecting our company's focus on the synergy between human expertise and machine efficiency.

We offer services that leverage digital agents that include AI-powered voice and chatbots, automated customer service tools, and analytics platforms that support human agents in delivering personalized customer interactions. Our solutions leverage artificial intelligence with humans in the loop to improve efficiency, reduce costs, and increase customer satisfaction by seamlessly integrating the capabilities of humans and machines.

Our approach typically focuses on industries where customer experience is a critical differentiator, such as retail, telecommunications, financial services, and healthcare. By leveraging technology to handle routine inquiries and transactions, human agents are freed to focus on more complex, sensitive, or high-value interactions where human empathy and understanding are paramount.

Our ethos centers around the belief that the future of effective customer sales and service lies in the intelligent blending of technology and human touch, ensuring that customers receive timely, accurate, and empathetic support at every touchpoint. We've served over two hundred successful clients and earned more than fifty awards for service and sales excellence.

One of them was a leading theme park and entertainment company—a US-based chain of marine mammal parks, oceanariums, and animal theme parks that are known for their zoological displays, aquarium exhibits, and live shows featuring various marine animals, including orcas, sea lions, dolphins, and other sea life.

Facing seasonal surges in customer calls and inquiries resulting in an increase in hold times, this company sought to improve customer experiences and provide AI self-service solutions while reducing costs. Despite the budget limitations, leveraging AI technology to upgrade the contact center was required. The goal was to provide customers with a convenient, personalized guest experience, allowing them to manage their inquiries more effectively. They wanted to achieve the following:

- **Improve customer service**, providing twenty-four-seven assistance to address guest queries.

- **Enhance user experience**, offering a more intuitive and efficient way for guests to engage.

- **Personalize service**, using AI to deliver personalized service.

- **Increase efficiency**, reducing the workload on human customer service representatives by automating responses to common inquiries.

In the words of the VP of contact center operations, "We needed an AI solution that could help resolve customer inquiries during peak season and a partner who understood how important our guests are to us, a partner who could enable us to make an emotional connection through our technology platform and drive families to our parks."

With these objectives in mind, Humach proposed a strategy in which digital AI agents would serve as the first point of contact for customers, and a significant percentage of customer issues could be addressed by providing answers to a handful of FAQs (park hours, ticket prices, events). These FAQs served as the foundation of the AI digital agent, interacting with guests to resolve their inquiries using 100 percent self-service AI digital agents.

Implementation included natural language processing (NLP) digital agents using advanced AI to understand customer queries and provide relevant, contextual answers based on access to real-time guest and company information. Application Programming Interfaces (APIs) between Humach and the company allowed access to customer account information, transaction history, and other data to provide personalized service. AI tuning and continuous learning allowed the AI digital agents to learn from interactions with users to improve their responses and functionality over time.

When AI digital agents were not able to complete the transaction using natural language processing and custom access to client data sources, guests were triaged to a contact center specialist for resolution, along with the complete transaction detail of AI digital agent experience so the guest would not have to repeat information. The conversation transcript assisted the contact center specialist in enhancing the guest experience and improving efficiency.

The benefits of this automation process were instantaneous. Using custom-developed AI digital agents and real-time access to client information systems, the AI digital agent was successful, delivering results in the first six months of operation. Humach handled 42 percent of this company's chats, and this company experienced:

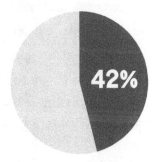

42% of Chats handled in the **first six months** of Humach's digital agent program.

- ♀ A total of **125,000** net new digital conversations

- ♀ A **700%** increase in intents handled over six months

- ♀ A reduced headcount by **52** FTEs (full-time equivalents)

Guest conversations were carefully monitored not only to increase response accuracy but also to identify new intents, sales opportunities, and customer pain points. From there, new conversation workflows were designed and implemented. Currently, the digital agent's most popular intents are those that were identified through this careful review process, not those initially provided during the discovery.

In addition to the digital AI deployment, Humach provided daily, weekly, and monthly reporting to boost contact center efficiency, quickly address problems, and identify opportunities for improvement. Now, every week the Humach digital agent handles thousands of chats without escalating to a live contact center specialist. To prepare for future information and technology updates, custom API connections to client data sources ensure that keeping answers up to date for all their parks remains as simple and efficient as possible.

Since the success of the AI digital agent deployment, the company has been exploring various technologies to enhance guest experiences and operational efficiencies across its parks. By leveraging the use of digital AI agents specifically for customer interactions in various aspects of its operations to improve efficiency and guest satisfaction, the company is now focused on continuous improvement in the guest experience, reducing both guest and employee effort.

Moorfields Eye Hospital and Google DeepMind

Another prominent example of how AI can enhance patient care with AI diagnostics involves the use of Google's DeepMind AI technology in partnership with Moorfields Eye Hospital in London. This collaboration between Moorfields Eye Hospital and Google DeepMind represents a landmark case study in using AI diagnostics to enhance patient care. In early 2024, Google DeepMind rebranded its AI services to Gemini.

As noted in chapter 1, DeepMind is renowned for its pioneering work in the field of AI, particularly in deep learning, neural networks, and reinforcement learning. DeepMind's mission is to solve intelligence, which it considers the most direct path to solve a wide array of other problems. Moorfield Eye Hospital is one of the leading and oldest eye hospitals globally. Led by Professor Sir Peng Tee Khaw, director of the National Institute for Health Research Biomedical Research Centre in Ophthalmology, this hospital began searching for how to help two million people in the UK who are living with sight loss, of whom around 360,000 are registered blind or partially sighted.

It is estimated that 98 percent of sight loss is caused by diabetes, and if the right treatment is applied at the right time, many cases can be prevented by early detection and treatment. This includes age-related macular degeneration and sight loss as a result of diabetes. As Professor Peng Tee Khaw stated,

> Our research with DeepMind has the potential to revolutionize the way professionals carry out eye tests and could lead to earlier detection and treatment of common eye diseases such as age-related macular degeneration. With sight loss predicted

to double by the year 2050 it is vital we explore the use of cutting-edge technology to prevent eye disease.[62]

The collaboration set out to explore how machine learning could help analyze eye scans and improve the treatment for patients with eye conditions with three objectives:

1. **Improve diagnostic accuracy:** Utilize AI to enhance the precision and efficiency of diagnosing eye diseases, such as diabetic retinopathy and age-related macular degeneration.

2. **Reduce diagnosis times:** Accelerate the process of diagnosing eye conditions to facilitate earlier treatment interventions.

3. **Support clinical decision-making:** Aid clinicians in decision-making processes by providing them with detailed analyses of eye scans.

Eye health professionals use scans of patients' eyes to detect and diagnose serious conditions and diseases, of which Moorfields completes more than three thousand optical coherence tomography (OCT) scans every week. The study was to use DeepMind-developed deep learning algorithms trained on thousands of historic depersonalized eye scans. These algorithms learned to identify signs of eye diseases. Then the AI system was rigorously validated in clinical environments to ensure its efficacy and safety in diagnosing conditions accurately.

Last came the close collaboration with clinicians and eye health professionals to ensure the AI system met clinical needs and could effectively integrate into existing workflows. The key thesis for the study was to determine whether AI technology could help improve the care of patients with sight-threatening diseases, such as age-related macular degeneration and diabetic eye disease, by making the analysis of OCT scans faster without losing any of the accuracy in diagnosis.

DeepMind used two types of neural networks (mathematical systems for identifying patterns in images or data), and immediately the AI system quickly learned to identify ten features of eye disease from highly complex OCT scans. The system was then able to recommend a referral decision based on the most urgent conditions detected, without any delay or accuracy degradation. Mustafa Suleyman, cofounder of DeepMind, stated,

> We set up DeepMind because we wanted to use AI to help solve some of society's biggest challenges, and diabetic retinopathy is the fastest growing cause of blindness worldwide. There are more than 350 million sufferers across the planet. Detecting eye diseases as early as possible gives patients the best possible chance of getting the right treatments. I really believe that one day this work will be a great benefit to patients across the NHS. We are proud of our NHS, and this is one of the ways I think we can help nurses and doctors continue to provide world-class care.[63]

To determine whether the AI system was making correct referrals based on the machine learning, clinicians also viewed the same OCT scans and made their own referral decisions based on their education and experience. The study concluded that AI was able to make the correct referral recommendation more than 94 percent of the time, closely matching the performance of expert clinicians.

AI also provided information that helped explain to eye care professionals how it arrived at its recommendations. This AI output included visuals of the features of eye disease it had identified on the OCT scan and the percentage level of confidence the system had in its recommendations. This functionality is crucial in helping clinicians scrutinize the technology's recommendations and check its accuracy before deciding the type of care and treatment a patient receives.

One of the initial challenges came down to data privacy, so all patient data was deidentified before being used for training the AI. The project also underwent strict ethical reviews and complied with all relevant data protection regulations.

Even with the success of this project and the AI results, the transition from research to clinical application requires careful planning, including regulatory approval, clinician training, and system integration into hospital IT environments. In the future, the AI system can be easily applied to different types of eye scanners, not just the specific model on which it was trained. Leveraging this could significantly increase the number of people who benefit from this technology and future-proof it, so it can still be used even as OCT scanners are upgraded or replaced over time.

Following the success of this collaboration, Moorfields and DeepMind aimed to explore further applications of AI in eye care, including predictive analytics for disease progression and personalized treatment plans. The partnership also set a precedent for how AI can be ethically and effectively integrated into healthcare settings.

By leveraging AI to improve the accuracy and efficiency of diagnosing eye diseases, this collaboration has paved the way for broader adoption of AI technologies in healthcare, demonstrating the potential to impact patient outcomes and clinical workflows significantly.

JPMorgan Chase COiN Platform

Another great example is JPMorgan Chase. This company is the behemoth of American banking, boasting a workforce exceeding three hundred thousand employees and catering to the financial needs of millions of customers nationwide. Within this colossal workforce are lawyers and loan officers, laboring through countless hours to decipher the intricacies of commercial loan agreements and legal documents.

In response to this monumental task, JPMorgan Chase unveiled the Contract Intelligence (COiN) platform, a revolutionary AI system designed to streamline document analysis. With the ability to swiftly review twelve thousand annual commercial loan agreements in mere seconds, COiN represents a quantum leap from the manual processes that would devour a staggering 360,000 hours of human effort.

Now COiN extends its capabilities beyond document interpretation to automate tasks like loan application analysis, liberating human employees to focus on more intricate and value-driven endeavors. Powered by an AI machine learning system and operating within a proprietary private cloud infrastructure, COiN marks a strategic leap forward in the bank's quest for operational excellence.

The inception of COiN heralded a strategic shift for JPMorgan Chase, aiming to enhance efficiency, slash operational costs, and bolster accuracy in document analysis. Leveraging image recognition and natural language processing (NLP), COiN transcends the limitations of human interpretation, extracting critical information and clauses from complex legal documents with precision and speed.

The platform continuously evolves through machine learning, adapting to new document formats and legal language nuances over time. Moreover, its seamless integration into JPMorgan Chase's existing infrastructure ensures a harmonious workflow between AI and human workers.

Following its implementation, COiN has proven to be a game changer, extracting approximately 150 relevant attributes from annual commercial loan and credit agreements within seconds, a feat that would otherwise demand hundreds of thousands of person-hours under manual review. Through pattern recognition and analysis, COiN empowers swift decision-making, minimizing errors and maximizing efficiency.

At the core of COiN's success lies JPMorgan Chase's foresight in investing in AI technology long before its widespread adoption. This strategic initiative exemplifies how the integration of artificial intelligence can revolutionize banking operations, enhancing efficiency, accuracy, and customer service while liberating human resources for strategic pursuits.

JPMorgan Chase's COiN platform stands as a beacon of innovation in the financial services industry, showcasing the transformative potential of AI. By automating the labor-intensive process of legal document review, the bank has not only achieved efficiency gains and cost savings but also elevated the accuracy of its operations.

As one commentator notes, "The ascent of AI and ML in the monetary business is changing the business scene. Expanding on the example of overcoming adversity of JPMorgan, it would bode well to relook at your hierarchical vision towards innovation and find a way to execute the more current innovation into your organization measures."[64]

This case study underscores the symbiotic relationship between human expertise and artificial intelligence in modern banking, highlighting AI's capacity to augment human capabilities and drive strategic initiatives forward. While financial institutions increasingly harness the power of AI and automation, they recognize the irreplaceable value of human experience in decision-making processes. And while AI can enhance analytical capabilities, human judgment remains pivotal in evaluating multifaceted factors influencing banking transactions.

The proliferation of AI and machine learning in banking heralds a new era of risk mitigation, fraud prevention, and enhanced customer service. COiN serves as a compelling testament to the fusion of human expertise and AI prowess, setting the stage for the future of banking.

Aravo Solutions

Aravo Solutions is another great example. This company combines AI and human expertise to enhance third-party risk management (TPRM). Unlike the futuristic AI depicted in movies like *Star Wars* or *2001: A Space Odyssey*, AI in TPRM doesn't mean relinquishing control to machines. Instead, it's about using AI to support and refine human intelligence, making decision-making processes more efficient and effective.

Michael Saracini, CEO of Aravo Solutions, said,

> Human beings are still responsible for teaching the AI to understand how decisions get made and what the tasks or outcomes should be by showing it examples. The advantage AI has over humans is the ability to efficiently analyze enormous amounts of data, quickly identifying patterns and trends without an explicit decision tree or model.

AI in TPRM is not about replacing humans but augmenting their capabilities to make quicker, more informed decisions. It's used for automating routine tasks, enhancing intelligence through data analysis, and predicting future risks by monitoring changes in third-party profiles. This approach allows TPRM teams to focus on tasks that require human insight, such as building relationships and conducting audits.

Automation in TPRM can range from simple business rules to advanced machine learning, where the system learns from past decisions to make new ones. Intelligence is where AI supplements human judgment with data-driven recommendations, and prediction involves AI monitoring for changes that could indicate a shift in a third party's risk profile.

The benefits of incorporating AI into TPRM include managing vast amounts of data more effectively, increasing confidence in decision-making, scalability, and ensuring adherence to organizational best practices. However, it's crucial to understand that AI is not a silver bullet. It cannot replace the need for a clear process, fix past mistakes, or compensate for poor data quality. Moreover, AI should not be seen as superior to human judgment.

Aravo Solutions advocates for a balanced approach to AI in TPRM, where AI supports human decision-makers rather than replacing them. This strategy ensures that TPRM programs can handle the complexity and volume of data involved, making them more efficient and less prone to error while still maintaining the human insight that is critical to effective risk management.

More Than a Trend

Through these case studies, we see that the integration of humans and machines is not just a passing trend but a fundamental paradigm shift reshaping how we approach business operations, problem-solving, and innovation. Across various industries, from healthcare to finance to customer service, the synergy between human intelligence and artificial intelligence has emerged as a powerful catalyst for driving progress and achieving unprecedented levels of efficiency and effectiveness.

At the heart of this integration lies a recognition of the unique strengths and capabilities that each brings to the table. On one hand, we have machines—AI and machine learning algorithms that possess unparalleled data-processing capabilities, speed, and scalability. These technological marvels excel at analyzing vast amounts of data, identifying patterns, and automating routine tasks with precision and efficiency. Their ability to crunch numbers, predict outcomes, and optimize processes has revolutionized industries, unlocking new possibilities for innovation and growth.

On the other hand, we have humans—with our creativity, emotional intelligence, intuition, and adaptability. Despite the remarkable advancements in AI technology, there are aspects of human cognition and behavior that remain elusive to machines. Our ability to understand context, interpret nuance, empathize with others, and exercise judgment in complex situations is what sets us apart and makes us indispensable in many domains. Moreover, humans possess a unique capacity for creativity and innovation—the ability to think outside the box, connect disparate ideas, and envision possibilities that have yet to be realized.

By integrating humans and machines, organizations can harness the complementary strengths of both to create synergistic partnerships that drive success. Whether it's in healthcare, where AI augments the diagnostic capabilities of medical professionals; in finance, where AI assists investors in making informed decisions; or in customer service, where AI enhances the quality of interactions between businesses and their clients, the benefits are clear. Together, humans and machines form dynamic teams that are greater than the sum of their parts, enabling organizations to achieve outcomes that would be unattainable by either alone.

Moreover, the integration of humans and machines holds profound implications for the future of work and society at large. Rather than viewing AI as a threat to human employment, we should embrace it as a tool for augmenting human capabilities and enhancing productivity. As machines take on more routine and repetitive tasks, humans are liberated to focus on higher-order activities that require creativity, critical thinking, and emotional intelligence. This not only leads to greater job satisfaction and fulfillment but also fosters a culture of innovation and continuous learning.

In conclusion, the integration of humans and machines represents a paradigm shift with far-reaching implications for business, society, and the future of work. By recognizing and leveraging the complementary strengths of both, organizations can unlock new levels of productivity, innovation, and success. However, realizing this vision requires a collaborative effort involving policymakers, business leaders, technologists, and educators to address the challenges and seize the opportunities that lie ahead. Only then can we truly harness the transformative power of human-machine integration to create a better, more prosperous future for all.

CHAPTER 10:

GUIDELINES FOR TRANSITIONING TO AN INTELLIGENT WORKFORCE

PICTURE A WORKPLACE WHERE HUMANS AND MACHINES team up, creating a synergy that leads to unprecedented levels of efficiency, innovation, and job satisfaction. It's not just a dream—it's the future.

As we embark on this journey, it's essential to recognize that navigating the transition to an intelligent workforce requires more than just flipping a switch. It demands strategic planning, ethical considerations, and a steadfast commitment to continuous learning and improvement.

Thus, we need clear objectives that align with our organizational vision and mission. After all, our goals should contribute to the broader success of the business and how the intelligent workforce fits into that equation. It's crucial to keep an eye on industry trends and benchmarks.

> ◉ What are the leading companies in our field doing?
>
> ◉ How are they integrating humans and machines for maximum efficiency and innovation?

By staying informed and proactive, we can ensure that our goals remain competitive and relevant.

Also, it's important to identify key areas within our organization that could use a little boost from AI and automation. Whether it's streamlining inefficient processes or uncovering insights hidden within our data, there's no shortage of opportunities to enhance our work with machine intelligence. As experts have noted,

> As the makeup of the modern workforce becomes more complex, two realities are emerging for organizational leaders: their organizations are relying more and more on external contributors like contractors, professional service companies, gig workers, crowdsourced contributors, app developers, and even certain technologies to achieve strategic goals and objectives. And their current workforce management systems aren't designed to bridge the management gap between internal and external contributors.[65]

To bridge this gap, we need to have some clear guidelines and action steps in place.

Step 1: Develop a Technology Strategy

Before you even think about shifting the culture in your organization, it's important to develop a clear technology strategy for your intelligent workforce that involves designing and integrating advanced tools and systems that enhance human capabilities with AI, machine learning, automation, and data analytics.

As one author notes, "Successful organizations share the common trait of relying on planning processes to ensure growth and operational efficiency. Just as with other business functions, every organization needs to have a considered and well-developed technology plan."[66] This process is not just about deploying cutting-edge technology but also about designing solutions that augment human skills, foster collaboration between humans and machines, and streamline workflows for greater efficiency and innovation.

This strategy should be rooted in a deep understanding of the current technological landscape and aligned with the organization's broader business goals of AI, automation, and digital tools to enhance human capabilities. A successful technology strategy for an intelligent workforce prioritizes scalability, flexibility, and security, ensuring that technological solutions can grow and adapt with the organization. It also involves identifying key areas where AI can make the most significant impact, such as improving operational efficiency, enabling better decision-making through data analytics, or enhancing customer experiences.

Critical to this strategy is the commitment to change management, fostering a culture that embraces innovation, continuous learning, and collaboration between humans and machines. By carefully planning and executing a technology strategy that addresses these elements, organizations can build a truly intelligent workforce that is prepared to meet the demands of the future.

To develop an effective strategy, identify your goals and create a detailed action plan. This plan should outline the necessary steps, required resources, and specific milestones to achieve those goals.

Step 2: Develop SMART Goals

When transitioning to an intelligent workforce, businesses can effectively navigate this evolution by utilizing SMART goals as guiding principles.

SMART goals, an acronym for Specific, Measurable, Achievable, Relevant, and Time-Bound,[67] provide a structured framework for setting objectives that are clear, actionable, and aligned with the organization's overarching vision. Let's break down how each element of SMART goals can be applied in the context of transitioning to an intelligent workforce.

First, SMART goals emphasize specificity, ensuring that objectives are clearly defined and unambiguous. When setting goals for the integration of humans and machines, businesses should articulate precisely what they aim to achieve. For example, rather than a vague goal like "improve efficiency," a specific goal might be "reduce processing time for customer inquiries by 20 percent through the implementation of AI-driven chatbots."

Measurability is the next key component of SMART goals, enabling businesses to track progress and evaluate success objectively. By establishing quantifiable metrics or key performance indicators (KPIs), organizations can monitor the impact of their efforts and make data-driven decisions. For instance, in the case of enhancing operational efficiency, measurable indicators could include average response time, error rates, or cost savings attributed to AI implementation.

Moreover, SMART goals emphasize achievability, ensuring that objectives are realistic and attainable within the organization's resources and constraints. While ambitious goals are commendable, setting unrealistic targets can lead to frustration and demotivation.

Businesses must assess their capabilities and allocate adequate resources to effectively support the transition to an intelligent workforce.

Relevance is another critical aspect of SMART goals, emphasizing the importance of aligning objectives with the broader strategic priorities and needs of the organization. Goals related to the integration of humans and machines should directly contribute to the organization's mission, vision, and long-term objectives. By ensuring relevance, businesses can prioritize initiatives that have the most significant impact on driving business success and competitiveness.

Finally, SMART goals highlight the importance of setting time-bound targets, establishing deadlines or milestones to create a sense of urgency and accountability. By defining specific timeframes for achieving objectives, businesses can maintain momentum, track progress, and course-correct if necessary. Time-bound goals also help prevent projects from languishing indefinitely and ensure that efforts remain focused and results-oriented.

In summary, by applying the SMART criteria to goal-setting, businesses can effectively navigate the transition to an intelligent workforce, setting clear, measurable, achievable, relevant, and time-bound objectives that drive success and innovation. By leveraging SMART goals as guidelines, organizations can maximize the potential of human-machine collaboration and position themselves for sustainable growth and competitiveness in the digital age.

Step 3: Collaborate and Create Clear Objectives

After you've established your goals, it's important to bring others into the process through collaborative, clear communication of your goals and objectives.

Effective communication channels are essential for facilitating collaboration across teams. This means your organization should implement tools such as project management software, instant messaging apps, and regular cross-functional meetings to facilitate easy and open communication. Encouraging transparency and promoting a culture of openness ensures that information is freely shared and updates are communicated regularly to keep everyone informed and engaged in the collaboration process.

I like the term *cross-functional collaboration*, which is a strategic approach that brings together team members from various departments and disciplines to work on a common project or toward a shared goal. This method leverages the diverse skills, perspectives, and expertise of individuals from different functional areas, such as marketing, finance, operations, IT, and human resources, to foster innovation, solve complex problems, and drive business growth more effectively.

Encouraging collaboration across all departments can break down silos, improve communication, and enhance understanding among different parts of the business. This collaborative environment enables the pooling of resources and ideas, leading to more creative solutions and a more cohesive strategy execution. Moreover, cross-functional collaboration is crucial in the context of deploying new technologies or strategies, such as AI and digital transformation initiatives, where the convergence of technical and domain-specific knowledge is essential for success.

By embracing cross-functional teamwork, your company can tap into a broader range of insights and capabilities, fostering a culture of continuous learning and adaptability that is vital in today's fast-paced business landscape. Encourage collaboration among IT, data science, and business units to ensure that AI initiatives align with business goals and are grounded in real operational needs. Employ change management strategies to manage the transition, addressing concerns and resistance from employees.

To ensure effective cross-functional collaboration, you must establish clear objectives and shared goals that align with your organization's overall vision. This involves ensuring that all team members and departments understand and buy into the shared vision, which helps prioritize tasks and resolve conflicts. By establishing clear, measurable goals that necessitate cooperation across functions, you can ensure your organization's teams are working toward the same outcomes, fostering cohesion and alignment.

Stakeholder engagement is a collaborative process that brings together a diverse array of individuals and groups to participate in the design of AI solutions. This includes end users, domain experts, ethicists, policymakers, and other stakeholders who may have a vested interest in the technology. By involving stakeholders throughout the design process, AI developers can gain valuable insights, perspectives, and feedback, ensuring that the resulting solutions are holistic, inclusive, and responsive to the needs and concerns of all involved parties.

Roles and responsibilities should be clearly defined to minimize overlap and gaps in responsibilities among team members. Each team member should understand their role, responsibilities, and how their work contributes to the overall project. Fostering accountability and a sense of ownership encourages team members to take ownership of their tasks and recognize the importance of their contributions. Additionally, appointing leaders or facilitators for cross-functional efforts provides guidance and support to drive collaboration effectively, while providing resources and support ensures that teams have the necessary tools, training, and expertise to work effectively across functions.

Step 4: Implement a Human-Centric Design

As you move toward embracing the intelligent workforce, the one thing you can't lose sight of is your focus on humanity. Having a human-centric design in AI means you place human needs, behaviors, and contexts at the forefront of the development process for artificial intelligence systems. This methodology emphasizes creating AI technologies that are not only technically advanced but also deeply integrated with human values, ethical considerations, and practical usability.

Implementing human-centric design in AI is about ensuring that artificial intelligence solutions are developed with a deep understanding of human needs, behaviors, and contexts at their core. This approach prioritizes the user experience, making AI systems more intuitive, accessible, and effective for the people who interact with them.

By adopting a human-centric design philosophy, developers and designers focus on creating AI technologies that augment human abilities without causing frustration or complexity. This involves engaging with end users throughout the development process to gather insights and feedback, thereby ensuring that the AI solutions are aligned with their actual needs and preferences.

It also means considering ethical implications, such as fairness, privacy, and transparency, to build trust and acceptance among users. Human-centric AI design emphasizes the seamless integration of AI into daily tasks, enhancing productivity and satisfaction without diminishing the human element. By putting humans at the heart of AI development, organizations can foster more meaningful interactions between humans and machines, leading to innovations that genuinely improve lives and work processes.

Understanding user needs and contexts is fundamental to developing AI solutions that truly resonate with users. At the heart of this endeavor lies empathetic design, which involves delving deeply into the lives of end users to comprehend their unique needs, challenges, and the environments they navigate. This profound understanding serves as the cornerstone for crafting AI solutions that are not only relevant and useful but also accessible to all who interact with them.

Moreover, embracing diverse user perspectives enriches the design process, ensuring that it captures the needs and experiences of a broad spectrum of users. By incorporating insights from individuals with varying backgrounds and experiences, AI solutions can be tailored to be inclusive, addressing the varied requirements of diverse populations and fostering a more equitable user experience.

Ethics by design is paramount in the development of AI systems, integrating ethical principles into every stage of the process. By prioritizing ethical considerations such as human rights, privacy, and dignity from the outset, AI developers can build systems that are not only technically proficient but also socially responsible, promoting trust and integrity in AI technologies.

As discussed earlier, transparency and accountability are central tenets in this endeavor, involving the design of AI algorithms and decision-making processes that are transparent and comprehensible to users. Providing visibility into how AI systems operate and fostering accountability ensures that developers and organizations are held responsible for the outcomes and implications of their AI systems, promoting ethical behavior and responsible use of AI technologies.

Prioritizing augmentation over automation underscores the development of AI systems that enhance human intelligence and capabilities rather than replacing human roles entirely. This approach fosters collaboration between humans and AI, recognizing that each brings unique strengths to the table. By augmenting human abilities with AI technologies, productivity and innovation can be amplified while preserving the human element in decision-making processes. User empowerment is central to effective AI design, involving the creation of AI tools that empower users by enhancing their capacity to make informed decisions, solve complex problems, and seize opportunities.

Accessibility and usability are fundamental considerations in ensuring that AI solutions are usable by people with disabilities, eliminating barriers to usage and interaction. By prioritizing accessibility and designing intuitive interfaces, AI developers can create inclusive experiences that enable all users to participate fully and independently in the digital world.

Ultimately, by embracing human-centric design principles, organizations can create AI technologies that not only deliver technical prowess but also align with human values, promoting trust, inclusivity, and empowerment in the digital age.

Step 5: Assess and Prepare Your Data

As you shift your organization toward the intelligent workforce, you will need to assess and prepare your data. This is essential, as the quality and organization of your data directly influence how well AI models perform and the results they deliver. Well-prepared data leads to more accurate and reliable analyses and predictions. To ensure your business has access to high-quality, relevant data for AI, there are several points to keep in mind:

POINT 1: CONDUCT A THOROUGH AUDIT OF ALL THE DATA WITHIN YOUR ORGANIZATION

This is a crucial first step in understanding and managing your data assets effectively. This audit involves systematically examining the sources, types, and storage locations of data across various departments and systems. To begin, it's essential to identify the sources of data within your organization. This includes both internal sources, such as databases, enterprise applications, and file systems, as well as external sources, such as third-party data providers or customer-generated data. By mapping out these sources, you can gain insights into the diversity and volume of data being generated and collected.

Next, it's important to classify the types of data present in your organization. Data can be broadly categorized as structured or unstructured. Structured data is organized in a predefined format with clear data types and relationships, often found in databases or spreadsheets. On the other hand, unstructured data refers to information that doesn't have a predefined data model or format, such as text documents, images, videos, or social media posts. Understanding the distribution and characteristics of structured and unstructured data helps in devising appropriate strategies for data management and analysis.

POINT 2: FOCUS ON THE QUALITY OF THE DATA

Ensure the data is accurate, complete, consistent, and relevant. Look out for missing information, duplicate records, or errors. It's also important to standardize the data to ensure consistency. This involves normalizing numerical values and standardizing formats like dates and times.

Continuously monitoring and improving data is crucial for maintaining data integrity and model accuracy. Regular monitoring of data quality and AI model performance helps detect anomalies and maintain data integrity. Iterative improvement involves refining data practices based on feedback from AI model performance and evolving business requirements, driving innovation and value creation.

By retaining relevant features and discarding irrelevant ones, we can enhance model performance. Feature creation generates new features from existing data to enrich the model's ability to recognize patterns and make accurate predictions. This may involve combining existing features or extracting meaningful features from unstructured data sources like text or images.

POINT 3: TRANSFORM THE DATA INTO A FORMAT THAT'S READY FOR AI MODELING

Transforming data into a format suitable for AI modeling involves several key steps to ensure compatibility and effectiveness. Initially, it's essential to preprocess the data, which includes handling missing values, scaling numerical features, and encoding categorical variables. Missing data can significantly impact model performance, so imputation techniques such as mean or median substitution or advanced methods like k-nearest neighbors (KNN) imputation are employed to fill in the gaps.

Additionally, numerical features often have different scales, which can skew the model's performance, so standardization or normalization techniques are applied to bring them onto a similar scale. Categorical variables, on the other hand, need to be encoded into numerical form using methods like one-hot encoding or label encoding, allowing the model to interpret them accurately.

Following preprocessing, feature engineering plays a pivotal role in shaping the data for AI modeling. Feature engineering involves creating new features or modifying existing ones to enhance the model's predictive capability. This process may include generating interaction terms between variables, extracting relevant information from text or images, or transforming variables to better align with the model's assumptions. By crafting informative features, the model gains a deeper understanding of the underlying patterns within the data, leading to improved predictive performance.

Feature engineering is often an iterative process, where domain knowledge and experimentation guide the creation of features tailored to the specific problem and dataset at hand. Ultimately, transforming the data through preprocessing and feature engineering ensures that it's primed and optimized for effective AI modeling, enabling the generation of meaningful insights and accurate predictions.

POINT 4: INTEGRATE AND CONSOLIDATE THE DATA

In addition to integrating and consolidating data, building a scalable data infrastructure is essential for organizations to effectively manage increasing data volumes and evolving processing needs. Scalability is crucial as businesses generate and collect more data over time. Cloud-based solutions offer scalability and flexibility, allowing organizations to adapt to changing demands seamlessly. By leveraging cloud services, businesses can dynamically scale their infrastructure resources, ensuring optimal performance and cost efficiency as their data needs grow.

Moreover, implementing data pipelines is instrumental in automating data workflows and streamlining the process of data collection, cleaning, transformation, and loading. Data pipelines enable organizations to maintain a continuous flow of updated and preprocessed data, ensuring that AI models have access to timely and reliable information for analysis and decision-making. By automating these data workflows, organizations can improve operational efficiency, reduce manual errors, and accelerate the development and deployment of AI applications, ultimately enhancing overall business agility and competitiveness in today's data-driven landscape.

POINT 5: ADDRESS DATA PRIVACY
AND COMPLIANCE

Following regulations like GDPR or HIPAA and handling personal or sensitive information with care is critical. Ensuring robust data security and governance measures is critical for organizations. Implementing comprehensive security measures helps safeguard the integrity and confidentiality of data assets. This entails setting up access controls to limit unauthorized access to sensitive data, employing encryption techniques to protect data in transit and at rest, and adhering to secure data storage protocols to prevent data breaches or unauthorized disclosures.

Furthermore, establishing a robust data governance framework is essential for effective data management and compliance. Data governance encompasses the formulation and enforcement of policies, standards, and procedures that govern data-related activities within the organization. This includes defining roles and responsibilities for data stewardship, ensuring adherence to data quality standards, and establishing protocols for data privacy and security.

By instituting a structured framework for data governance, organizations can foster transparency, accountability, and compliance with regulatory requirements, thereby mitigating risks and safeguarding against potential data breaches or legal liabilities.

Step 6: Leverage Expertise
and Keep Evolving

Leveraging external expertise is a pivotal element in crafting a technology strategy for deploying AI and fostering an intelligent workforce. This approach allows organizations to tap into specialized knowledge and insights that can accelerate the development and implementation of AI initiatives.

By engaging with AI research institutions, technology consultants, and industry-specific AI solution providers, companies can gain access to cutting-edge AI developments, best practices, and lessons learned from other implementations. This external expertise can help identify the most suitable AI technologies and approaches for the organization's unique needs, navigate the complexities of AI integration, and avoid common pitfalls. Furthermore, partnering with external experts can enhance the organization's ability to train and support its workforce in adopting new AI tools, ensuring a smoother transition to an intelligent workforce.

Consider partnerships with AI vendors, universities, and research institutions to access expertise and resources that can accelerate AI initiatives, as it opens up opportunities for collaboration on innovative projects, potentially leading to unique AI applications that can provide a competitive edge. Engage with the wider AI ecosystem, including startups, to stay abreast of emerging technologies and approaches.

An interdisciplinary approach is essential for creating AI systems that are truly human-centric. This approach integrates knowledge and expertise from various fields, such as psychology, sociology, design, and computer science, to inform the development of AI technologies. By drawing on insights from diverse disciplines, AI developers can gain a deeper understanding of human behavior, preferences, and needs, allowing them to design systems that are not only technically proficient but also culturally sensitive, socially responsible, and aligned with human values and aspirations. Ultimately, an interdisciplinary approach fosters innovation, creativity, and collaboration, driving the creation of AI systems that enhance the human experience and contribute positively to society.

Ultimately, leveraging external expertise ensures that the technology strategy is not only aligned with the latest AI advancements but is also pragmatically tailored to deliver tangible business outcomes and empower the workforce.

FOSTER A CULTURE OF INNOVATION AND LEARNING

Encourage a culture of continuous learning and adaptation among employees to keep pace with technological advancements. As automation and AI become more prevalent, there will be a growing need for employees with advanced technological skills. This could lead to a skill gap if current employees do not have these skills and new ones are not found or trained.

Invest in training programs to upskill employees in AI and related technologies, ensuring they can work effectively alongside intelligent systems. Invest in the technologies and skill sets of today, but also those of tomorrow. This involves staying updated with emerging trends and technologies. Be diligent about ensuring employees are adequately trained to work with AI and automation technologies. This may involve providing technical training for IT staff and training customer service representatives and other employees on how to use the new tools effectively.

Your business must adapt its strategies and culture to embrace these changes. This includes changing workflow design and workspace design to adapt to an era where people work more closely with machines. If automation results in a significant reduction in employment or greater pressure on wages, some ideas such as conditional transfers, support for mobility, universal basic income, and adapted social safety nets could be considered and tested.

It's not just about technical skills, such as data analysis or machine learning, but also about fostering adaptability, critical thinking, and an understanding of ethical AI use. Your organization must create comprehensive learning programs that include workshops, online courses, and hands-on projects to offer practical experience with AI tools.

Encouraging a culture of continuous learning and curiosity is essential, as it motivates your employees to embrace AI innovations and explore new ways to apply them in their work. By investing in upskilling initiatives, companies can unlock the full potential of their workforce, ensuring that employees are not only prepared to use AI but also to innovate and lead in an AI-driven future.

This approach not only enhances operational efficiency and innovation but also empowers employees, boosting their confidence and career resilience in the face of technological change.

MONITOR, MEASURE, AND ADAPT

Monitoring and measuring the success of an intelligent workforce project involves establishing clear, relevant metrics that align with the project's objectives and the organization's broader goals. Key performance indicators (KPIs) might include productivity enhancements, cost savings, error reduction rates, employee engagement and satisfaction levels, customer satisfaction scores, and the speed of decision-making processes.

Implementing a robust analytics framework is essential to track these metrics in real time, enabling timely adjustments. Data should be gathered not only on the performance of AI and automation technologies but also on how these tools impact human work processes and outcomes. Qualitative feedback from employees and customers can provide insights into the user experience and the effectiveness of human-AI collaboration.

Regular review meetings should be scheduled to assess progress against goals, discuss challenges, and identify opportunities for improvement. This holistic approach ensures that the project remains aligned with strategic objectives, maximizes the return on investment, and truly enhances the capabilities of the intelligent workforce. Be prepared to adapt AI strategies based on performance data and evolving business needs.

Your business should gauge the success of transitioning to a smarter workforce by monitoring various metrics. It should track the work completed by the intelligent workforce, analyzing key indicators like financial savings, capacity utilization, and revenue increases. This provides valuable insights into operational efficiency and areas for improvement.

Employee engagement and satisfaction serve as crucial markers of a successful transition. These are assessed through surveys, feedback sessions, and turnover rates, reflecting the morale and commitment of the workforce. Similarly, increased customer engagement resulting from data-driven strategies signals success. Customer surveys, net promoter scores, and retention rates offer tangible evidence of enhanced customer interaction.

Productivity metrics such as active user numbers and overall productivity levels provide further validation of the transition's success. Efficiency in training and onboarding processes, measured by the speed at which new hires become proficient, reflects the effectiveness of these strategies. Moreover, the rate at which employees adapt to new technologies and workflows is indicative of a successful transition. Performance reviews and feedback sessions help assess this aspect.

KPIs play a crucial role in evaluating the impact of AI and automation on these metrics. They are complemented by qualitative assessments, including employee feedback and customer reviews, which provide insights into the effectiveness of human-AI collaboration and the overall user experience. Regular analysis of these metrics offers a comprehensive view of the intelligent workforce's performance, pinpointing areas of success and opportunities for further optimization.

This approach ensures that integrating AI into the workforce not only achieves technical objectives but also enriches the organization's human capital, fostering sustainable growth and maintaining a competitive edge.

Step 7: Keep Focused on Customer and Employee Experience

In your pursuit of an intelligent workforce, where humans seamlessly collaborate with advanced technologies like AI and automation, you have a golden opportunity to transform both customer and employee experiences. By placing customer experience at the forefront, you can harness AI to deliver personalized and efficient services while empowering your employees through automation to tackle more strategic and fulfilling tasks.

When it comes to enhancing customer experience, AI becomes your most valuable asset for analyzing vast amounts of customer data to uncover insights and preferences. With AI-driven analytics, you can tailor your products, services, and interactions to meet individual customer needs, ultimately boosting satisfaction and fostering loyalty. Implementing chatbots and virtual assistants ensures round-the-clock customer support, providing timely assistance and making customers feel valued and supported at any time of day.

Equipping your employees with AI-driven tools not only streamlines their workflow but also enhances their ability to cater to customer needs effectively. With insights derived from AI analysis of customer data, your staff can engage in more informed and empathetic interactions. Moreover, AI predictions regarding customer trends enable proactive problem-solving and opportunity identification, allowing you to stay ahead of the curve and consistently exceed customer expectations.

On the employee experience front, AI and automation play a pivotal role in alleviating the burden of repetitive tasks, allowing your employees to focus on higher-value work that nurtures creativity and strategic thinking. Personalized learning platforms powered by AI facilitate continuous professional development tailored to individual growth areas, fostering a culture of learning and career advancement. Additionally, AI-supported flexible working arrangements promote work-life balance by enabling effective collaboration and communication, regardless of physical location or schedule.

By utilizing AI to monitor employee well-being and offer personalized recommendations for stress management and health initiatives, you demonstrate a commitment to creating a supportive and nurturing workplace environment. These strategies collectively contribute to a positive organizational culture that values both customer satisfaction and employee well-being, ultimately driving success and prosperity for your entire organization.

Step 8: Prioritize Ethical AI Use

While this point was the emphasis of chapter 7, it's worth reiterating here. When you're navigating the topic of ethical AI use, it's crucial to understand the principles and practices that ensure artificial intelligence technologies are developed and deployed in a fair, transparent, and accountable manner while also respecting human rights and privacy.

As AI systems continue to exert influence over various aspects of society and individual lives, ethical considerations become paramount to prevent biases, discrimination, and potential harm. The ultimate goal is to create AI algorithms that are free from prejudices, thereby avoiding unfair treatment of certain groups. Transparency is key, ensuring that individuals understand how AI systems make decisions and can challenge them if adversely impacted.

Moreover, ethical AI use necessitates strict adherence to data privacy laws, ensuring that personal information is handled responsibly and with consent. Organizations and developers must also consider the broader societal impacts of AI, such as potential job displacement and the digital divide. Prioritizing ethical AI use enables us to leverage technology's benefits to enhance human welfare, foster trust in AI systems, and mitigate risks that could undermine public confidence and the technology's potential for good.

To ensure ethical AI use, you must adopt clear guidelines that encompass fairness, transparency, accountability, and the impact on employment. Additionally, the implementation of AI and automation raises legal considerations. It's essential for businesses to ensure compliance with all relevant laws and regulations, using these technologies ethically and transparently.

Building trust involves safeguarding data privacy and security and effectively communicating how AI is used to benefit both customers and employees. By following ethical guidelines, you can build trust with stakeholders and society, ensuring that AI technologies are used responsibly and for the greater good.

Step 9: Start Small and Scale Gradually

As you embark on your AI journey, consider implementing AI pilots, also known as small-scale projects designed to test and evaluate the feasibility, performance, and impact of AI solutions before wider deployment. These pilots serve as a crucial step in your AI integration process, offering you the opportunity to explore the capabilities of AI technologies in a controlled environment with minimal risk.

By focusing on specific use cases or business processes, AI pilots enable you to gather valuable insights on how AI can enhance your operational efficiency, improve decision-making, and create innovative services or products. Through these projects, you can identify potential challenges, such as data quality issues, integration complexities, or user adoption hurdles, and address them effectively before scaling up.

Moreover, AI pilots provide an opportunity for your stakeholders to experience firsthand the benefits of AI, building support and enthusiasm for broader AI initiatives within your organization. Importantly, these projects help fine-tune your AI models, ensuring they are aligned with your organizational goals and ethical standards and laying a solid foundation for successful AI adoption across the enterprise.

Begin with a small, manageable project to test your AI solutions in controlled environments, allowing for adjustments before full-scale implementation. Pilot projects enable you to test AI and automation solutions, assess their impact, and learn from the experience. Measure the success of these pilot projects against your objectives, using the insights gained to refine your approach and resolve any issues before scaling up.

Once you have identified successful initiatives through pilot projects, gradually scale your AI and automation initiatives across the organization. Plan for broader integration, fostering a culture of continuous improvement where feedback is regularly sought and used to enhance your AI solutions and workforce collaboration.

Scaling an AI solution after a successful pilot project begins with a thorough evaluation of the pilot outcomes, analyzing outcomes, identifying best practices, and understanding any challenges faced. Ensure that your infrastructure, both technical and organizational, is prepared to support wider deployment by upgrading hardware and software as necessary, ensuring data quality and availability, and integrating the AI solution with existing systems and workflows.

Assess how well the project met its objectives, including the performance of the AI solution and its impact on business metrics. This involves gathering quantitative data from performance metrics and qualitative feedback from stakeholders involved in or affected by the pilot. It's crucial to identify what worked well, any challenges encountered, and how the technology was adopted by the workforce, including technical challenges, data quality issues, team collaboration, and stakeholder engagement.

Lessons learned from the pilot should inform the development of best practices for future AI integration efforts. Additionally, you should evaluate the pilot's scalability, considering technical infrastructure, data readiness, and workforce adaptability for broader deployment. Strategic planning for future projects should also take into account emerging AI trends and potential impacts on your organization's industry. By systematically reviewing pilot performance and leveraging insights gained, you can refine your approach to building an intelligent workforce, ensuring future projects are more effectively executed and aligned with long-term business goals.

You can then utilize the success of the project to create a strategic road map for AI within your organization, identifying new areas where AI can add value, such as operational efficiency, customer experience, or innovation. Conduct a thorough assessment of current capabilities, data readiness, and any gaps in technology or skills. Engage stakeholders across the organization to ensure buy-in and align AI goals with departmental needs, and develop a phased plan that starts with achievable pilot projects to demonstrate quick wins, learn, and build momentum.

Each phase should outline specific objectives, technologies to be deployed, necessary investments in infrastructure and talent, and metrics for measuring success. Incorporate continuous learning and adaptability into the road map, allowing for adjustments based on technological advancements and lessons learned from initial projects. A well-constructed strategic road map serves as a guide, ensuring that AI initiatives are purposeful, integrated into the organizational fabric, and positioned to drive long-term competitive advantage.

Finally, leverage the project's success to secure funding and resources for future AI initiatives, demonstrating the return on investment to decision-makers. Following a successful AI project, the focus should be on leveraging the momentum to drive further innovation, improve operational efficiencies, and create additional value for the organization. This includes scaling successful solutions, fostering a culture of continuous learning and improvement, and strategically planning for future AI initiatives.

By following these steps, you can clearly identify actionable goals for integrating an intelligent workforce, aiming to enhance efficiency, productivity, innovation, and employee satisfaction, all while delivering superior value to customers.

A Strategic Imperative

As you conclude this chapter, you've embarked on a journey toward a future where the synergy between humans and advanced technologies like AI is not just a possibility but a strategic imperative for success in the modern business landscape.

As you prepare to implement these guidelines within your own organization, remember that the transition to an intelligent workforce is not a one-time event but an ongoing process of evolution and refinement. Embrace the opportunity to learn from each step along the way, celebrating successes and embracing failures as valuable learning experiences.

Above all, keep in mind that the ultimate goal of transitioning to an intelligent workforce is not just to drive efficiency or increase profitability but to create meaningful and impactful outcomes for both your employees and your customers. By staying true to your organizational values and putting people at the center of your AI initiatives, you'll be well-positioned to thrive in an increasingly digitized and interconnected world.

With careful planning, thoughtful execution, and a commitment to continuous improvement, you'll be well on your way to building a future where humans and machines work together seamlessly to achieve extraordinary results.

CONCLUSION

IN A DANCE, YOU AND YOUR PARTNER BRING YOUR OWN strengths, skills, and style to the collaboration. Similarly, humans and machines each contribute their unique abilities and expertise to achieve a shared objective. Just as dancers must synchronize their movements, you and the machines must coordinate your actions to perform tasks efficiently and effectively. Like skilled dance partners, you complement each other's strengths and compensate for weaknesses, resulting in a graceful and seamless performance that captivates and achieves the desired outcome.

As you conclude your journey through the landscape of the intelligent workforce, you've witnessed the intricate interplay between humans and machines, each contributing their unique strengths and navigating their inherent vulnerabilities. Together, you redefine the very essence of productivity, creativity, and innovation, shaping a future where collaboration between humans and machines is the norm rather than the exception.

Throughout this exploration, I've done my best to provide a road map for understanding and navigating this new terrain, where AI and human intelligence merge to create boundless opportunities for businesses and individuals alike. As you step into a future where AI and humans coexist and collaborate, it's imperative to grasp the profound implications of this shift.

In today's digital era, characterized by rapid technological innovation and disruption, leaders like you are confronted with an unprecedented array of opportunities and challenges. The dynamism of this landscape demands a proactive approach, where adaptability becomes not just a desirable trait but a fundamental requirement for success. As a leader, you must be agile in your response to change, leveraging emerging technologies to stay ahead of the curve and drive innovation within your organization.

However, this journey is not without its obstacles. Human nature tends to resist change, often clinging to the comfort of familiar routines and methodologies. Overcoming these inherent tendencies toward complacency and resistance requires courage, vision, and a commitment to embracing new possibilities. In an evolving world where the only constant is change, leaders who can navigate this delicate balance between tradition and transformation will emerge as catalysts for growth and pioneers of innovation.

Your organization will face the pivotal task of harmonizing the relationship between humans and technology, ensuring that advancements empower rather than replace the human workforce. If you navigate this balance well, your business will emerge as a frontrunner in an increasingly competitive landscape.

The vision of a blended future, where human and machine capabilities seamlessly converge, holds the promise of reshaping every facet of your existence. From augmented decision-making in the workplace to enhanced daily experiences in your personal lives, the symbiotic relationship between humans and technology is poised to unlock unprecedented levels of efficiency, creativity, and innovation.

> *As we weave closer ties with technology, we step into a future where our connections transcend intellect, suggesting the dawn of a shared consciousness. This evolution blurs the boundaries of our individual selves, proposing a harmony between us and technology that could redefine our essence. It invites us to ponder: Which parts of ourselves are we prepared to leave behind? And what do we want to become?"*

♀ DENNIS WAKABAYASHI,
Chief executive officer, The Global Voice of CX

In this future, workplaces will undergo a transformative metamorphosis to accommodate seamless collaboration between you and the machines. Augmented by AI-driven insights, you will be liberated from mundane tasks, allowing you to focus on unleashing your creative potential and driving groundbreaking innovation across industries.

Beyond the confines of the workplace, smart technology will enrich your daily experiences, enhancing accessibility, inclusivity, and sustainability. Businesses will revolutionize customer experiences through hyper-personalization and seamless interactions, while ethical considerations will remain paramount in the development and use of AI systems.

As you navigate this evolving landscape, human values such as empathy, ethics, and creativity will emerge as guiding beacons. In a world where technology is omnipresent, fostering a culture of lifelong learning and adaptability will be imperative, ensuring you are equipped with the skills to thrive in a rapidly changing environment.

The journey toward a blended future is not without its challenges, but it's also teeming with opportunities for progress and innovation. By embracing the symbiosis between technological capabilities and human values, you can forge a path toward a more efficient, creative, and inclusive world. The future is here, where humans and machines collaborate harmoniously to create a truly intelligent workforce. In the end, the most remarkable outcomes will be achieved when people like you and machines work hand in hand, pushing the boundaries of possibility.

And now, as you conclude your journey, I'll leave you with one final question: *Was this book written by a human or machine?* Based on the core principles we've outlined, I think you probably know the answer: Human + Machine = Humach.

CONTACT

TIM HOULNE | HUMACH

5 Cowboys Way, Suite 300

Frisco, Texas 75034

Office: (972) 694-0600

Email: thoulne@humach.com

ENDNOTES

CHAPTER 1

1   Deloitte, "The Digital Workplace," Deloitte, accessed October 6, 2023, https://www2.deloitte.com/content/dam/Deloitte/be/Documents/technology/The_digital_workplace_Deloitte.pdf.

2   PricewaterhouseCoopers (PwC), "Workforce of the Future," PwC, accessed October 6, 2023, https://www.pwc.com/gx/en/services/workforce/publications/workforce-of-the-future.html.

3   Matthew Boyle, "Dimon Says Remote Work 'Doesn't Work' for Younger Staff, Management," Bloomberg, January 19, 2023, https://www.bloomberg.com/news/articles/2023-01-19/jamie-dimon-says-work-from-home-doesn-t-work-for-young-staff-management?embedded-checkout=true.

4   Shayna Waltower, "Is Working from Home More Productive? The Surprising Answer," Business News Daily, accessed October 6, 2023, https://www.businessnewsdaily.com/15259-working-from-home-more-productive.html.

5   "How Many Phones Are in the World?" BankMyCell, accessed October 6, 2023, https://www.bankmycell.com/blog/how-many-phones-are-in-the-world.

6   "Workforce of the Future: The Competing Forces Shaping 2030," TR2050, accessed October 6, 2023, https://www.tr2050.com/insights/workforce-of-the-future-the-competing-forces-shaping-2030/#:~:text=We%20are%20living%20through%20a,organisation%20must%20seek%20from%20staff, 3.

7   Elaine Pofeldt, "Are We Ready for a Workforce That Is 50% Freelance?," Forbes, October 17, 2017, https://www.forbes.com/sites/elainepofeldt/2017/10/17/are-we-ready-for-a-workforce-that-is-50-freelance/?sh=2938f7553f82.

8   "Strategies for the Gig Workforce," Deloitte, accessed October 6, 2023, https://www2.deloitte.com/us/en/blog/human-capital-blog/2021/strategies-for-the-gig-workforce.html.

9   "Strategies for the Gig Workforce," Deloitte.

10   "Workforce of the Future: The Competing Forces Shaping 2030," TR2050.

11   "Global Survey: The State of AI in 2020," McKinsey & Company, accessed October 6, 2023, https://www.mckinsey.com/business-functions/quantumblack/our-insights/global-survey-the-state-of-ai-in-2020.

12   John Seabrook, "Has the Pandemic Transformed the Office Forever?," The New Yorker, accessed October 6, 2023, https://www.newyorker.com/magazine/2021/02/01/has-the-pandemic-transformed-the-office-forever?utm_campaign=falcon&utm_social-type=owned&utm_source=twitter&utm_brand=tny&utm_medium=social&mbid=social_twitter.

13   "Global Survey: The State of AI in 2020," McKinsey & Company.

14   "Global Survey: The State of AI in 2020," McKinsey & Company.

15   "Global Survey: The State of AI in 2020," McKinsey & Company.

16 Dan Anderson, "Elon's Imagining the Internet Center: Pew Research Center Release New Study on Artificial Intelligence," Elon University News, accessed October 6, 2023, https://www.elon.edu/u/news/2018/12/09/elons-imagining-the-internet-center-pew-research-center-release-new-study-on-artificial-intelligence/.

17 https://deepmind.google/technologies/alphago/#:~:text=As%20simple%20as%20the%20rules,times%20more%20complex%20than%20chess.

18 Google DeepMind, "AlphaGo: The Movie, Full Award-Winning Documentary," YouTube video, posted March 13, 2020, https://www.youtube.com/watch?v=WXuK6gekU1Y&ab_channel=DeepMind.

19 Garry Kasparov, "Condolences to Lee Se-dol on losing game one to AlphaGo. I hope he can recover, but the writing is on the wall," Twitter, March 9, 2016, https://twitter.com/kasparov63/status/707532443541225472?lang=en.

20 Google DeepMind, "AlphaGo."

21 Google DeepMind, "AlphaGo."

22 Andrew Kokes, "The Future of Excellent CX: Leveraging Enhanced Generative AI," Forbes, June 26, 2023, https://www.forbes.com/sites/forbescommunicationscouncil/2023/06/26/the-future-of-excellent-cx-leveraging-enhanced-generative-ai/?sh=1c8820331d98.

23 "Ep. 6: Kasparov," Ten Words Podcast, accessed October 6, 2023, https://tenwords.libsyn.com/ep-6-kasparov.

24 "Workforce of the Future: The Competing Forces Shaping 2030," TR2050.

25 Academy of Management Journal 59, no. 3, (2016): 737, http://dx.doi.org/10.5465/amj.2016.4003.

CHAPTER 2

26 https://technologymagazine.com/data-and-data-analytics/history-digital-transformation.

27 Need citation

28 Tim Houlne and Terri Maxwell, The New World of Work Second Edition: The Cube, The Cloud and What's Next (Charleston: Advantage Media Group), 23.

29 European Investment Bank, "Digitalisation: Developing Countries," accessed February 13, 2024. https://www.eib.org/en/stories/digitalisation-developing-countries.

30 https://www.ibm.com/topics/internet-of-things.

31 "Digitizing Isn't the Same as Digital Transformation," Harvard Business Review, March 2021, https://hbr.org/2021/03/digitizing-isnt-the-same-as-digital-transformation.

CHAPTER 3

32 https://www-formal.stanford.edu/jmc/whatisai.pdf.

33 Google AI. "Discover Generative AI," accessed February 13, 2024, https://ai.google/discover/generativeai/.

34 "Machine Learning Explained," MIT Sloan Ideas Made to Matter, November 1, 2023, https://mitsloan.mit.edu/ideas-made-to-matter/machine-learning-explained].

35 https://www.ibm.com/topics/natural-language-processing.

36 https://enterprisersproject.com/article/2019/5/rpa-robotic-process-automation-how-explain.

37 https://www.pewresearch.org/internet/2018/12/10/improvements-ahead-how-humans-and-ai-might-evolve-together-in-the-next-decade/.

CHAPTER 4

38 Pew Research Center, "How Americans Think About Artificial Intelligence," Pew Research Center: Internet, Science & Tech, March 17, 2022, https://www.pewresearch.org/internet/2022/03/17/how-americans-think-about-artificial-intelligence/.

39 Forbes Tech Council, "The Future of the Workforce: How Human-AI Collaboration Will Redefine the Industry," Forbes, May 4, 2023, https://www.forbes.com/sites/forbestechcouncil/2023/05/04/the-future-of-the-workforce-how-human-ai-collaboration-will-redefine-the-industry/?sh=3802d7a37983.

CHAPTER 5

40 David De Cremer and Garry Kasparov, "AI Should Augment Human Intelligence, Not Replace It," Harvard Business Review, March 2021, https://hbr.org/2021/03/ai-should-augment-human-intelligence-not-replace-it.

41 Brandon Goleman, Emotional Intelligence: For a Better Life, Success at Work, and Happier Relationships, Improve Your Social Skills, Emotional Agility and Discover Why it Can Matter More Than IQ (EQ 2.0), (2019), 11.

42 Adam Waytz, "Milken Conference: AI, Workers, Ethics," CNN, May 1, 2019, https://www.cnn.com/2019/05/01/perspectives/milken-conference-ai-workers-ethics/index.html.

CHAPTER 6

43 Goldman Sachs, "Artificial Intelligence," accessed December 20, 2023, https://www.goldmansachs.com/intelligence/artificial-intelligence/index.html.

44 Ashley Welch, "Artificial Intelligence is Helping Revolutionize Healthcare as We Know it," September 13, 2023, https://www.jnj.com/innovation/artificial-intelligence-in-healthcare?&utm_

source=google&utm_medium=cpc&utm_campaign=GO-
USA-ENG-PS-Corporate+Equity-GP-EX-RN-NB_STORIES_
ARTIFICIAL+INTELLIGENCE&utm_content=AI+-
+Articles&utm_term=ai+in+healthcare+article&gclid=CjwKCAiAv-
oqsBhB9EiwA9XTWGQCWS4h0tdykHxhqRgLnE2fLVg3QXnB
WMXTaxlM2NqASiDA7sQPoRhoCtuUQAvD_BwE&gclsrc=aw.ds.

45   "Rise of the Chatbots: How AI Changed Customer Service," Salesforce, accessed December 20, 2023, https://www.salesforce.com/ap/
hub/service/how-ai-changed-customer-service/#:~:text=Customer%20
service%20agents%20can%20realistically,handle%20multiple%20
queries%20at%20once.

46   Gina Schaefer and Ryan Sanders, "Humans with Machines: Harnessing the Power of Intelligent Automation (IA) by Effectively Managing
a Mixed Workforce of People and Machines," July 16, 2020, https://
www2.deloitte.com/us/en/blog/human-capital-blog/2020/humans-
with-machines.html.

CHAPTER 7

47   Christina Pazzanese, "Ethical Concerns Mount as AI Takes Bigger Decision-
Making Role," Harvard Gazette, October 2020, https://news.harvard.
edu/gazette/story/2020/10/ethical-concerns-mount-as-ai-takes-bigger-
decision-making-role/.

48   UNESCO, "Artificial Intelligence: Examples of Ethical Dilemmas,"
April 2023, https://www.unesco.org/en/artificial-intelligence/
recommendation-ethics/cases.

49   Pazzanese, "Ethical Concerns Mount as AI Takes Bigger Decision-
Making Role."

50   Pazzanese, "Ethical Concerns Mount as AI Takes Bigger Decision-
Making Role."

51   Pazzanese, "Ethical Concerns Mount as AI Takes Bigger Decision-
Making Role."

52   Jackie Shoback, "Data Privacy vs. Data Security: Four Implications for Business Leaders," Forbes, January 9, 2023, https://www.forbes.com/sites/forbesbusinesscouncil/2023/01/09/data-privacy-vs-data-security-four-implications-for-business-leaders/?sh=ec75dca6afae.

53   Deloitte, "Bringing Transparency and Ethics into AI," Deloitte, accessed February 8, 2024, https://www2.deloitte.com/content/dam/Deloitte/nl/Documents/innovatie/deloitte-nl-innovation-bringing-transparency-and-ethics-into-ai.pdf.

54   Deloitte, "Bringing Transparency and Ethics into AI."

55   Alonzo Martinez, "Balancing Innovation and Compliance: Navigating the Legal Landscape of AI in Employment Decisions," Forbes, October 31, 2023, https://www.forbes.com/sites/alonzomartinez/2023/10/31/balancing-innovation-and-compliance-navigating-the-legal-landscape-of-ai-in-employment-decisions/?sh=6d2eed7d2da2.

CHAPTER 8

56   Kweilin Ellingrud et al., "Generative AI and the Future of Work in America," McKinsey, accessed February 8, 2024, https://www.mckinsey.com/mgi/our-research/generative-ai-and-the-future-of-work-in-america.

57   Ian Bird, "Reskilling Your Workforce in the Time of AI," IBM, December 6, 2023, https://www.ibm.com/blog/reskilling-your-workforce-in-the-time-of-ai/.

58   Aditya Malik, "How New Rules and Regulations Can Help Create an Ethical Corporate Workspace With the Growth of AI," Forbes, July 17, 2023, https://www.forbes.com/sites/forbestechcouncil/2023/07/17/how-new-rules-and-regulations-can-help-create-an-ethical-corporate-workspace-with-the-growth-of-ai/?sh=312e58206eb7.

59   Expert Panel, "15 Significant Ways Quantum Computing Could Soon Impact Society," Forbes, April 18, 2023, https://www.forbes.com/sites/forbestechcouncil/2023/04/18/15-significant-ways-quantum-computing-could-soon-impact-society/?sh=308b9209648b.

60 Ellingrud et al., "Generative AI and the Future of Work in America."

61 Ellingrud et al., "Generative AI and the Future of Work in America."

CHAPTER 9

62 Google, "Moorfields Venture Eyes AI to Speed Diagnosis with OCT," Ophthalmology Innovation Summit, accessed February 8, 2024, https://ois.net/google-moorfields-venture-eyes-ai-to-speed-diagnosis-with-oct/.

63 Mustafa Suleyman, "Excited to Announce a New Medical Research Partnership with DeepMind Health," Moorfields, accessed February 8, 2024, https://www.moorfields.nhs.uk/content/excited-announce-new-medical-research-partnership-deepmind-health.

64 "JPMorgan COIN: A Case Study of AI in Finance," Superior Data Science, accessed February 9, 2024, https://superiordatascience.com/jp-morgan-coin-a-case-study-of-ai-in-finance/.

CHAPTER 10

65 Elizabeth J. Altman, Robin Jones, and Sue Cantrell, "Managing the extended and connected workforce: A framework for orchestrating workforce ecosystems," Deloitte Insights, April 2023, https://www2.deloitte.com/us/en/insights/topics/talent/contingent-workforce-solutions.html?id=us:2ps:3gl:difow24:awa:di:011024:workforce%20planning:b:c:kwd-10772441&gad_source=1&gclid=CjwKCAiAt5euBhB9EiwAdkXWO1t9lN7iIqGrYHYoDiJ4Gz_s1SE4lc3SqvxLmUINlDLVM57xAlWhuxoCNXIQAvD_BwE.

66 OSIbeyond, "Technology Strategy 101," OSIbeyond (blog), accessed February 8, 2024, https://www.osibeyond.com/blog/technology-strategy-101/.

67 George T. Doran, "There's a S.M.A.R.T. Way to Write Management's Goals and Objectives," Management Review 70, no. 11 (1981): 35.

www.ingramcontent.com/pod-product-compliance
Lightning Source LLC
La Vergne TN
LVHW092008050326
832904LV00017B/317/J